CURRY, Alan. Under the microscope, by Alan Curry, Robin F. Grayson, Geoffrey R. Hosey. Van Nostrand Reinhold, 1983 (c1982). 160p ill bibl index 82-19973. 14.50 ISBN 0-442-21598-3. CIP

Designed for those who know very little about the microscope and its use but who are seriously interested in learning about microscopy. The reader is taken through the choice and use of the microscope, constructing a microscope, techniques for mounting and staining specimens, collecting and studying aquatic organisms, microscopic study of minerals, photomicrography, special methods such as dark field microscopy, phase-contrast, interference and fluorescence microscopy, and finally the design and use of both the transmission and scanning microscope. The book, written in very straightforward nontechnical language, is exceptionally well illustrated with photographs and drawings to illustrate techniques and instruments. Informative material about specimens is also presented. Two useful appendixes on materials and suppliers; bibliography of some 50 references. Should serve as a useful source of basic information for high school and college teachers and students, as well as for the serious amateur microscopist.

Under the Microscope

To Our Parents

Under the Microscope

Alan Curry
Robin F. Grayson
Geoffrey R. Hosey

VNR VAN NOSTRAND REINHOLD COMPANY
New York Cincinnati Toronto London Melbourne

Library of Congress Catalog Card Number 82-19973
ISBN 0-442-21598-3

Printed in the U.S.A.

Published in the United States of America by
Van Nostrand Reinhold Company Inc.
135 West 50th Street
New York, New York 10020

14 13 12 11 10 9 8 7 6 5 4 3 2 1

Library of Congress Cataloging in Publication Data

Curry, Alan.
Under the microscope.

Bibliography: p.
Includes index.
1. Microscope and microscopy. I. Grayson, Robin F.
II. Hosey, Geoffrey R. III. Title.
QH205.2.C87 1983 502'.8'2 82-19973
ISBN 0-442-21598-3

Contents

Preface

This book is an attempt to show that having few resources and lack of professional training need not limit those interested in microscopy. Many people may feel that microscopy is too complicated a subject and that good quality microscopes are too expensive to buy, but we hope to show that this is not the case. This book is a practical guide for those engaged in laboratory work and for those to whom microscopy is a fascinating hobby, but the enormous contribution to scientific research made by the microscope (including advanced electron microscopes) is also described.

Acknowledgements

We would like to thank the following for their invaluable assistance in the preparation of this book: Mrs C. Anfield, Mr R. Benson, Dr R. Butler, Mr and Mrs J. T. Curry, Dr P. Evennett, Mrs J. Grayson, Mrs S. Hosey, Mr J. Ironmonger, Mr D. Lewis, Mr B. Madden, Mr W. Richardson, Dr D. Sigee, Mr L. Simmons and Mr R. Stacey.

We are also grateful to the following for their help: Dr I. Burton, Mr S. Chapman, Mr J. Cockerlin, Dr R. Earll, Mr D. Edwardson, Dr M. Harris, Dr P. Hasleton, Mr J. Hutton, Mr B. Jolley, Dr J. Kennaugh, Mr C. Lister, Mr P. McCormick, Mrs G. McMaster, Mr P. Matthews, Miss J. Roberts, Mr D. Roberts-Jones, Mr R. Shuttleworth, Mr P. Sullivan, Mr D. Wilkinson and Miss S. Wilson.

We would like to express our appreciation to the following companies for their assistance: A. Gallenkamp & Co. Ltd (Olympus), The Projectina Co. Ltd (Nikon), Reichert-Jung UK, and Vickers Instruments.

Our special thanks are due to the typist, Mrs Audrey Simmons, for her hard work and her patience with us.

Additional thanks to Oxford Illustrators for preparation of the diagrams and to GJK Limited for page layouts.

Finally, for the use of laboratory facilities, we would like to thank Dr D. M. Jones, Director of the Public Health Laboratory, Withington Hospital; Bolton Institute of Technology; and Wigan College of Technology.

Introduction

From its simple beginnings in the seventeenth century the microscope has evolved into a complex and sophisticated instrument. Its use has become a discipline in its own right, with its own terminology and methods; and as an item of laboratory equipment the microscope is virtually indispensable.

Early microscopes bear little resemblance to those available today. They consisted of a lens arranged to view objects impaled on a pin. Their modern descendants are the hand lens and the dissecting microscope, which are suitable for observing whole objects, but which do not achieve high magnifications.

The beginning of the eighteenth century saw an innovation which led to a great increase in the popularity of the microscope – the use of light transmitted through the specimen, for viewing items mounted upon a suitable slide. This is the principle upon which the modern compound light microscope works, and most of the procedures involved in microscopy are to do with preparing the specimen in a form suitable for such examination.

The twentieth century has seen the development of a fundamentally different sort of microscope, which uses electrons rather than light for producing an image. The electron microscope can resolve features that are much smaller than those resolvable in the light microscope, because the wavelength of an electron beam is much shorter than that of light (table 1).

Although the microscope has contributed to all branches of science, its impact has probably been greatest in biology and medicine. In the nineteenth century, the light microscope promoted the science of histology (the study of tissues), just as today the electron microscope has promoted the study of cell ultrastructure. Methods of sectioning and staining tissues were developed in the nineteenth century, and new insights into the anatomy of living organisms resulted. Today these techniques have a new importance, in the diagnosis of disease by the removal for examination of a small sample of tissue (a biopsy) at operation.

However, the use of the microscope is by no means restricted to the examination of tissues. Many whole organisms and inorganic materials such as rock sections can be investigated with the microscope, and techniques for doing so are described in this book.

The preparation of specimens for microscopic examination involves the use of materials which are not always readily available. Many of these can

one METRE (m)	=	1000 MILLIMETRES (mm)
one MILLIMETRE (mm)	=	1000 MICROMETRES (μm)
limit of resolution of the human eye	=	100 micrometres (μm) approx.
diameter of a bacterium	=	1 micrometre (μm) approx.
one MICROMETRE* (μm)	=	1000 NANOMETRES* (nm)
red light	=	740 to 620 nanometres (nm) wavelength
orange light	=	620 to 585 nanometres (nm) wavelength
yellow light	=	585 to 575 nanometres (nm) wavelength
green light	=	575 to 500 nanometres (nm) wavelength
blue light	=	500 to 445 nanometres (nm) wavelength
indigo light	=	445 to 426 nanometres (nm) wavelength
violet light	=	426 to 390 nanometres (nm) wavelength
ultraviolet light	=	less than 390 nanometres (nm) wavelength
limit of resolution of the light microscope	=	250 nanometres (nm) approx.
diameter of a polio virus	=	26 nanometres (nm)
width of a DNA molecule	=	2 nanometres (nm)
wavelength of an electron beam at 10,000 volts acceleration	=	0·012 nanometres (nm)
wavelength of an electron beam at 60,000 volts acceleration	=	0·005 nanometres (nm)

Table 1. SI units of measurement used in microscopy. * The micrometre (μm) was formerly known as the MICRON (μ). The nanometre (nm) was formerly known as the MILLIMICRON (mμ). The ANGSTROM (Å) is a redundant unit, equal to 0·1 nanometre (nm).

be obtained from a specialist supplier, but the private individual may have to use alternatives where necessary (pointed out in the text). In most cases stains and fixatives can be made up from basic ingredients or purchased ready made. Methods for preparing these are given in Appendix I, and materials referred to in the text should be sought there. Appendix II lists a number of suppliers of the materials to be used. Many of the chemicals are, of course, either toxic or caustic and should therefore be treated with great care. Some of the techniques described, such as sectioning, involve additional hazards, and these will be indicated where appropriate.

Units of measurements used in this book are SI (*Système International*) which are now accepted as standard. Inches are given in brackets where appropriate. In microscopy, the basic units of length in the SI system are the millimetre, micrometre and nanometre; and their relationships to one another are shown in table 1.

The first consideration for the individual interested in microscopy is to acquire the use of a microscope. The features and adjustments which determine the capabilities of microscopes are discussed in Chapter 1. Microscopes vary in complexity, and in quality, and this is often reflected in the purchase price which may reach many thousands of pounds. Although a few models are within the price range of a private individual, it is perfectly possible to construct a home-made microscope, for which the most expensive items are the objectives and the eyepiece. These must be

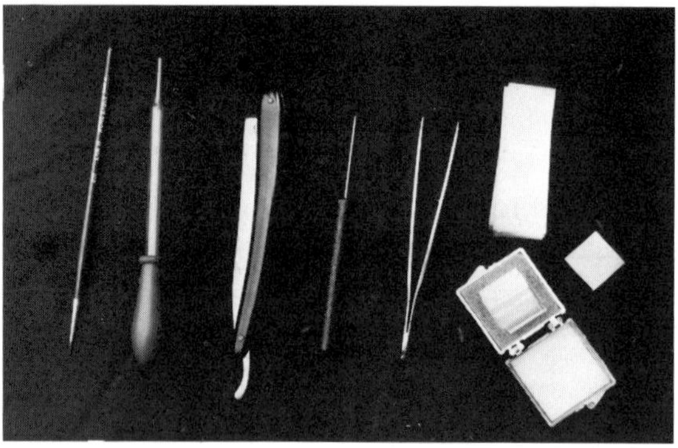

Fig. 1 Instruments necessary for microscopy. From left to right: artist's fine brush, bulb pipette, razor, mounted needle, narrow forceps, slides and coverslips.

purchased separately. Details of the construction of such a microscope are given in Chapter 2.

The next stage is the preparation of specimens for examination under the microscope. The instruments required are shown in fig 1. Many items may be mounted with very little prior treatment. Examples of these, and general methods of mounting, are described in Chapter 3. More complex preparative techniques are described in Chapter 4. Chapter 5 is concerned with the examination of living organisms and Chapter 6 with the preparation and examination of geological specimens. The observation of living aquatic organisms has a long history, dating back to the seventeenth century, when hay infusions were first prepared and 'infusoria' discovered in the preparations. The techniques used here are relatively straightforward, whereas those required for the examination of geological specimens are unique.

Having prepared and examined specimens on slides, microscopists may wish to make permanent records of their material, and this can be achieved by drawing or through photomicrography. Chapter 7 shows how the camera may be used to record any interesting features seen through the microscope.

Some specimens are particularly difficult to examine under the microscope, and various special techniques are available to extract the information from such specimens. A selection of these, including some which can be adapted for use with less complex microscopes, are described in Chapter 8.

The last two chapters deal with the electron microscope in its various forms and show some of the uses to which these instruments are normally put. Although the electron microscope has allowed great advances to be made in our knowledge of cells, micro-organisms and viruses, it has by no means made the light microscope redundant or obsolete. Indeed, they are complementary instruments.

A bibliography is included for those who wish to obtain further information about the specimens, the techniques or the instruments.

1

Choosing and Using a Microscope

Choosing a microscope is like choosing a camera, since both are optical systems ranging from relatively cheap, fairly simple models to expensive, highly sophisticated professional instruments. Microscopy, like photography, can be pursued as a hobby or as a career. The difference is that in photography the beginner has expert advice readily available from photographic shops, clubs and magazines, whereas this sort of help is rarely available for those interested in microscopy.

Before purchasing your own microscope, it is worth enquiring at a local school or college to see if you can try a good student microscope. This will provide you with experience of the qualities, capabilities and limitations of these instruments, and foster a more critical attitude towards the optical and mechanical features of available microscopes. Advertising claims such as 'magnifies 1000 times', and irrelevant attributes such as zoom magnification systems can then be treated with due scepticism, and preferably avoided. It is worth considering the purchase of good quality lenses from a specialist company, in order to build your own microscope, as described in Chapter 2.

Sometimes a microscope can be obtained second hand from a school, university or reputable dealer, and these older instruments are often a better buy than many cheap new microscopes. However, since an old microscope may have been abused, or have parts missing, it is necessary to check the condition of its lenses and mechanical controls. A few manufacturers of high quality research microscopes also offer good value, low priced instruments at a price comparable to a good single lens reflex 35 mm camera. In fact some of these manufacturers also make cameras, so their names may be familiar.

Projector as a Microscope

To start microscopy a microscope is not necessarily essential. If a colour transparency projector is available it can be used to magnify certain objects up to 50 times. To begin with, try a few strands of wool mounted between the two pieces of glass of a transparency holder. Such double glass transparency mounts can be purchased from photographic shops.

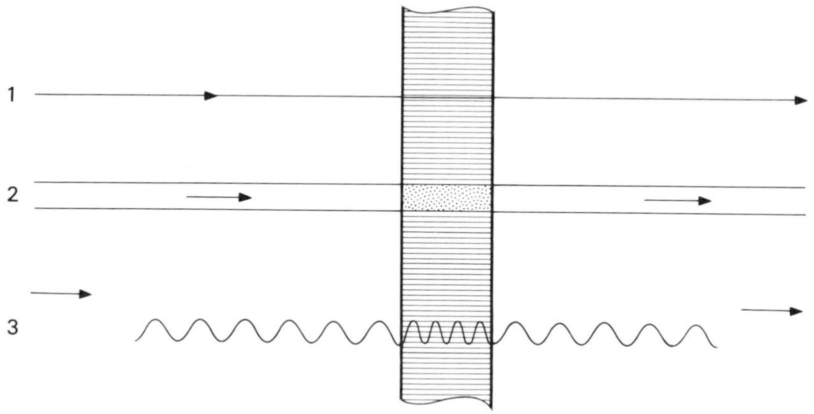

Fig. 2 Behaviour of light passing through a glass sheet at right angles. This can be portrayed as a ray (1), a beam (2) or a wave train (3). Notice that the light is slowed down within the glass.

Trimmed acetate peels of human fingerprints and fossil corals can be projected using this method (see Chapters 3 and 5). In addition, specimens already mounted on glass microscope slides can be projected. However, because transparency mounts used in 35 mm projectors measure 50 mm × 50 mm (2 in. × 2 in.) most glass microscope slides will not fit, being 76 mm × 25 mm (3 in. × 1 in.). Trim 12 mm ($\frac{1}{2}$ in.) or more off each end of a microscope slide with a glass cutter. Now mount the shortened slide in a cardboard transparency mount of standard 50 mm (2 in.) square size. This can then be inserted in the projector and viewed on a screen. To check magnification just project a portion of a transparent plastic ruler trimmed to fit.

The great advantage of a projector over a microscope is that large area specimens can be observed. It is an ideal device for audience participation in microscopy. Its only serious drawback is the low magnification that can be achieved.

Simple Microscopes and Single Lenses

Lenses are of two basic types. Those that converge or concentrate light are known as positive lenses and are thicker in the centre than at the edges. Those that diverge or spread light are known as negative lenses and are thinner in the centre than at the edges. The human eye contains a positive bi-convex lens which, unlike a glass lens, can be squeezed or stretched to alter its curvature. This means it can focus as easily upon a near as on a distant object.

Simple microscopes consist of a single lens only, or a single group of lenses. Generally, such a lens is called a *hand lens*, for the term *simple microscope* has faded from everyday use.

How can a single lens manage to magnify a small object? Although light can be thought of as travelling in waves, it can also be thought of as travelling in the form of rays; thus light from the object travels outwards in all directions in straight lines at a steady speed through the air. Until, that is, rays enter a substance that has a higher optical density than the air. The light rays temporarily slow down until they pass through the object of higher optical density and escape back into the air, where their original higher speed is resumed. Glass is a good example of an optically dense substance. If a horizontal light beam hits a vertical sheet of *plane* glass its speed will be slowed down while passing through the glass, but the beam will continue along in its original direction undeviated (fig. 2). If, however, it should hit the glass at any other angle the beam will be bent, both upon entering the glass and upon leaving (fig. 3). This is due to one part of the incident beam striking the glass and slowing down before another part of the same beam strikes the glass. This has the effect of bending the beam as it enters the glass. An analogy may be helpful here. Think of a bulldozer, which moves on two tracks. If both tracks move at the same speed, then the bulldozer will travel in a straight line. If, however, one track slows down, then the bulldozer will turn in the direction of that track. This bending phenomenon in an optically dense substance is

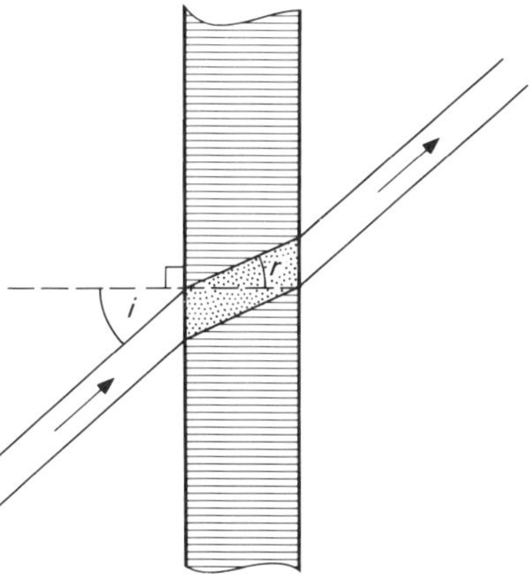

Fig. 3 A beam of light obliquely striking a glass plate is bent (refracted). *r* is the angle of refraction and *i* the angle of incidence.

termed *refraction*. Notice that, as the angle of incidence (i) increases, so too does the angle of refraction (r). Close study reveals a mathematical relationship of great value in optics:

$$\frac{\text{sine of the angle of incidence}}{\text{sine of the angle of refraction}} = \text{a constant value}$$

This constant value for light travelling from air into an optically dense substance, is known as the *refractive index* of that substance.

The eye contains a lens which exploits refraction to focus an image of any object on the retina at the back of the eye (fig. 4). Lenses put reality upside down and back to front. The brain interprets the retinal image so that we see an object the right way round. However, the brain cannot reorientate the image produced by microscope lenses, so this is reversed and upside down.

A person with perfect vision can only distinguish, ie resolve, features greater than 0·1 mm in size. This is because the retina of the eye has a limited number of light sensitive cells (especially cones) receiving the image of the object. Birds of prey, which are dependent upon hunting have, by evolution, packed more cones into the retina and so have much better definition in the image produced at the back of the eye.

The microscope and the hand lens work by putting a bigger retinal image upon the back of the eye, so more nerve endings can respond to it.

Compound Microscopes

The simple microscope consists of a single group of lenses (or even just a

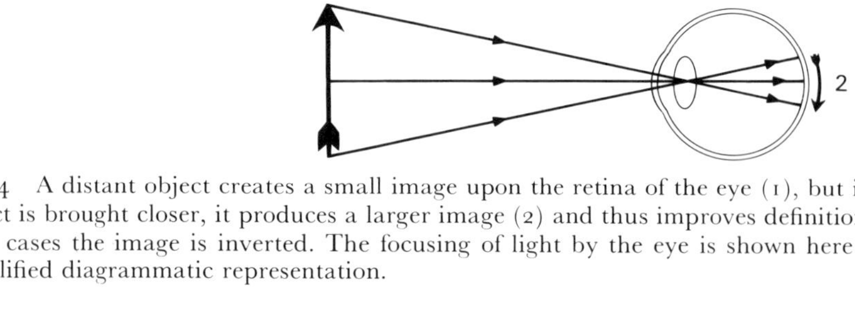

Fig. 4 A distant object creates a small image upon the retina of the eye (1), but if the object is brought closer, it produces a larger image (2) and thus improves definition. In both cases the image is inverted. The focusing of light by the eye is shown here as a simplified diagrammatic representation.

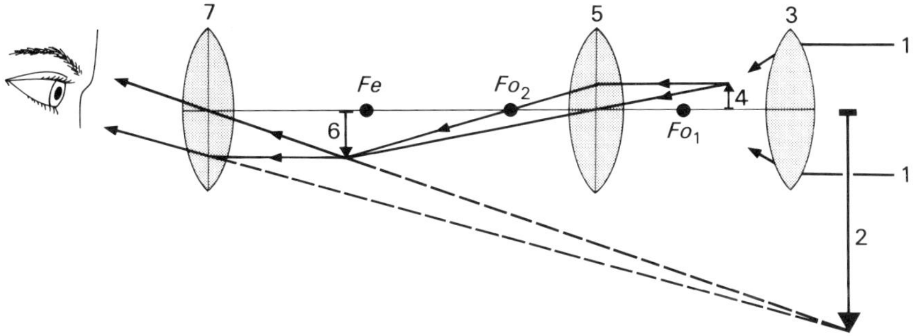

Fig. 5 Light path through compound microscope, showing how image is magnified in two stages, first by objective and then by eyepiece. Fo_1 is the front focal length of the objective lens; Fo_2 is the rear focal length (back focal plane) of the objective; Fe is the focal length of the eyepiece. (1) Light path. (2) Final magnified image. (3) Condenser. (4) Object. (5) Objective. (6) Primary image. (7) Eyepiece. Refraction shown from central plane of lens for convenience only; rays refract from lens surface.

single lens). Such simplicity restricts the useful magnification of this microscope to about 20 times (×20). This limitation in magnification is overcome by a more complex arrangement known as the *compound microscope.*

The compound microscope consists of two main lens groups which magnify the object in two stages (fig. 5). The first group, known as the *objective*, produces the primary image, which is further magnified by the second group, known as the *eyepiece* or *ocular*. The objective and eyepiece consist of two or more individual lenses.

A typical compound microscope with its major parts labelled is shown in fig. 6. This type is called a monocular microscope because light passes through a single eyepiece. Greater ease of working is possible with a binocular microscope, such as that illustrated in fig. 7 which enables both eyes to be used simultaneously.

MAGNIFICATION IN COMPOUND MICROSCOPES

The total magnification of a compound microscope can be calculated by multiplying the magnifying power of the eyepiece (say ×10) by the magnifying power of the objective (say ×40), which gives the overall magnification of ×400 in this example. Both the eyepiece and objective usually have their respective magnifying powers engraved upon them. The calculation is based upon the assumption that the eyepiece and objective, when in focus, are the distance apart intended by the manufacturer. This distance is known as the *mechanical tube length* (MTL) or *body tube length* and is the distance between the shoulder of the objective and the

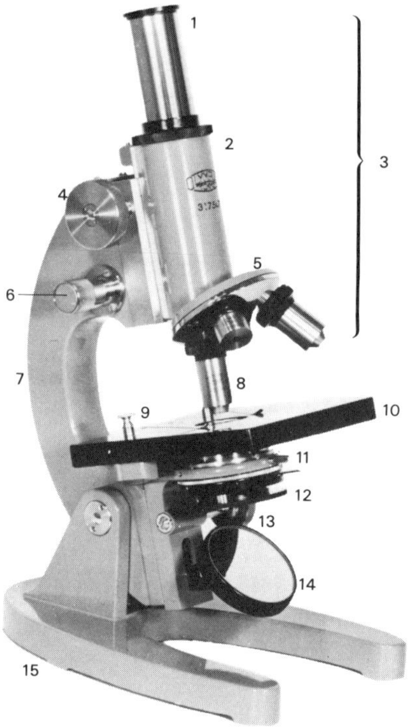

Fig. 6 A typical student microscope, the Olympus MJA 500F. (1) Eyepiece. (2) Body tube. (3) Mechanical tube length. (4) Coarse focusing control. (5) Triple nosepiece. (6) Fine focusing control. (7) Stand. (8) Objective. (9) Slide retaining clip. (10) Stage. (11) Iris diaphragm control. (12) Substage condenser. (13) Condenser control. (14) Mirror. (15) Foot.

flange of the eyepiece, as shown in fig. 6. When using lenses intended for a different tube length, the new magnification can be calculated using the following simple equation:

$$\text{magnification} = \frac{\text{mechanical tube length (mm)}}{\text{focal length of objective (mm)}} \times \text{eyepiece magnification}$$

The two commonest tube lengths used nowadays are 160 mm and 170 mm.

The Objective

The objective is the most important part of the optical system of a compound microscope (fig. 8). It provides the primary image with its resolved fine detail and much of the ultimate magnification seen by the eye

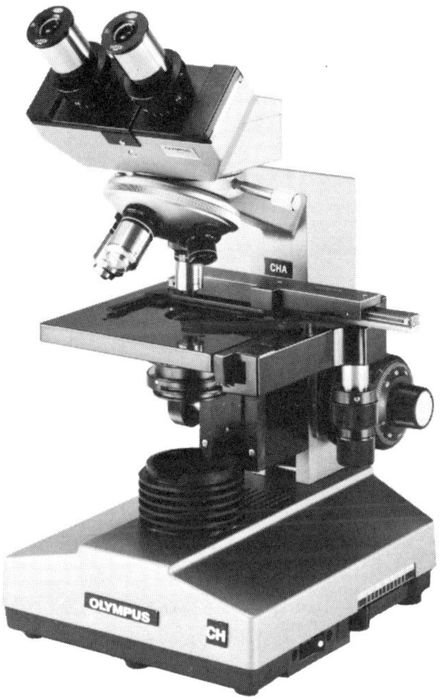

Fig. 7 A binocular microscope with built-in illumination and mechanical stage, the Olympus CHA.

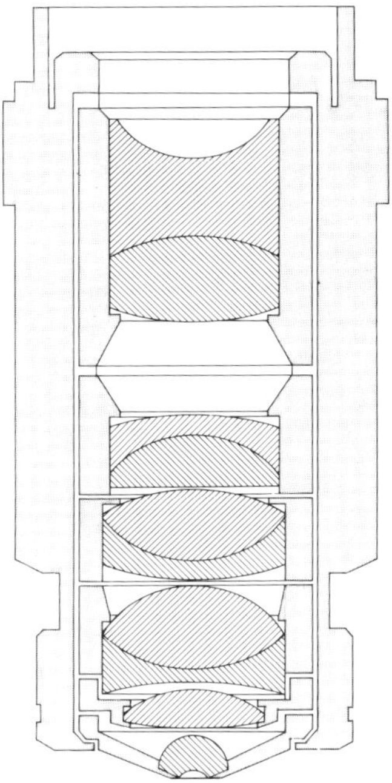

Fig. 8 Section through a Vickers ×63 Apoplan (plan apochromatic) objective, showing the complex arrangement of individual lenses.

(fig. 9). No single objective fulfils all the needs of a microscopist, so several different objectives are used. These can be mounted on a revolving nosepiece which allows objectives to be clicked into place as required. The objectives carried on such a nosepiece vary in length, the shortest being of lowest magnification and the longest of highest magnification. If one objective of a matched set from a single manufacturer is focused on to a specimen, then the rest will be roughly in focus, when each in turn is swung into position. Such a set of lenses is said to be *parfocal*.

The clearance between the lowest point of the objective and the specimen in focus is called the *working distance*. The working distance becomes smaller as the magnification of the objective lens increases (see table 2 for features of objective lenses). It is obvious that there is very little margin for error with high magnification objectives and great care must be exercised to avoid 'ramming' the objective into the specimen. Most high magnification objective lenses made these days are spring-loaded so that accidental touching of objective and specimen does not result in damage

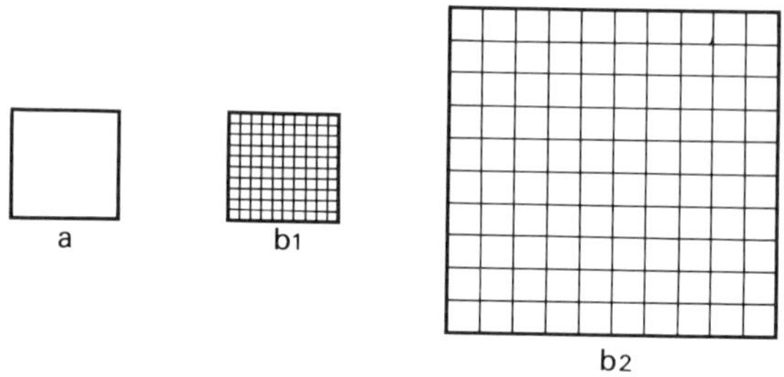

Fig. 9 The difference between resolution and magnification. (a) is a square with no discernible internal detail. Improved resolution reveals internal detail (b1). By magnifying (×3) no new detail is added – but the detail present can be seen in larger form (b2). For the compound microscope the limit of resolution (R) in μm is given by the formula

$$R = \frac{0.61\,\lambda}{\text{NA}}$$

where λ is the wavelength of the light used, and NA the numerical aperture of the objective.

to either. It is a good practice to find and focus the specimen with a low magnification objective first and then switch to higher magnification. When you finish observing the specimen, again switch to a low magnification objective before removing the specimen slide. This reduces the possibility of damaging a high magnification objective with its short working distance.

Objectives from major manufacturers have a standard screwthread so that the lenses can be swapped from one microscope to another, regardless of maker. However, some cheap microscopes have objectives with non-standard threads. Such lenses cannot be used on any other make of microscope.

Objectives used in microscopes are of several types and they usually have markings engraved on them (fig. 10). The basic markings applied to most objective lenses are magnification and *numerical aperture* (NA) of the lens. Older objectives will have the focal length given instead of the magnification.

magnification	focal length	numerical aperture	working distance
×4	40 mm	0·10 NA	10 mm
×10	16 mm	0·17 to 0·30 NA	4 to 8 mm
×20	8 mm	0·4 to 0·45 NA	1·2 to 5·7 mm
×40	4 mm	0·65 to 0·85 NA	0·2 to 0·68 mm
×100 (oil immersion)	2 mm	1·2 to 1·3 NA	0·11 to 0·16 mm

Table 2. Some characteristics of objectives. (Assuming the microscope tube length is 160 mm.)

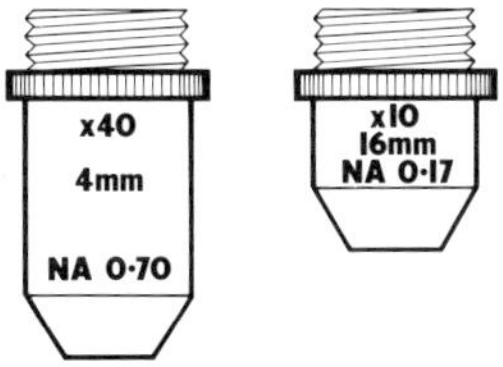

Fig. 10 Two objectives of different magnification, showing the information normally engraved upon them.

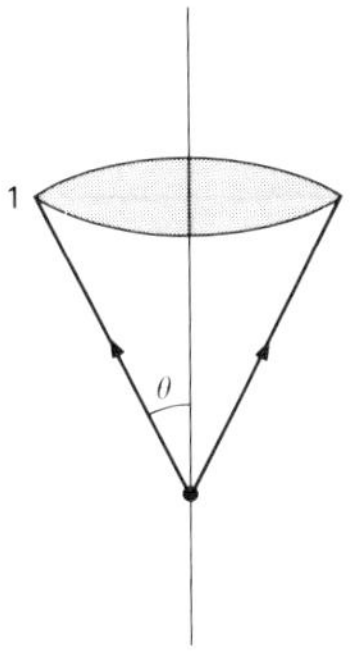

Fig. 11 The numerical aperture of a lens $NA = n \sin \theta$, where n is the refractive index of the medium between the coverslip and the objective, and θ is half the maximum angle of acceptance of the objective lens (1).

NUMERICAL APERTURE (NA)

An objective lens must allow one to distinguish between structural details set very close together in the specimen. The ability of an objective to resolve fine detail is expressed in terms of numerical aperture (NA). When an objective is used normally, with air between it and the specimen (a 'dry' objective lens), the maximum possible numerical aperture of that lens is 1 (fig. 11). In practice, the upper limit of a dry objective lens is about 0·95. Objectives of higher magnification and resolving power are designed to work with a special oil (immersion oil – refractive index about 1·5) between the specimen and the first lens element. These *oil immersion objectives* have a maximum numerical aperture of 1·5. The higher the numerical aperture of a lens the better it resolves fine detail. For our purposes the numerical aperture of an objective is related to the diameter (aperture) of the lens and its focal length. Numerical aperture can be described as half the angle of acceptance of the lens.

CHROMATIC ABERRATION

Light is composed of colours (red, orange, yellow, green, blue, indigo and violet), which combine to produce the 'white' light we see. Each colour represents a different wavelength of light. A single lens does not bring the rays of these colours to the same focus (fig. 12). Blue light, which has a shorter wavelength than red light, is brought to focus before red light because it is refracted (or bent) more strongly by the lens. This defect in a lens is called *chromatic aberration* and gives rise to unpleasant colour haloes around the specimen. This problem can be much improved by combining lenses made of different types of glass, so that it is possible to bring rays of two colours to a common focus and obtain a reasonable compromise for the other colours in white light.

Fig. 12 Chromatic aberration of a lens. The lens refracts the shorter wavelength blue light (1) more than it refracts the longer wavelength red light (2).

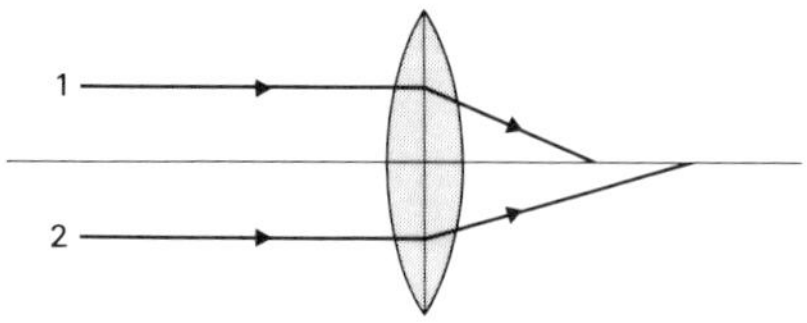

SPHERICAL ABERRATION

Combining lens elements also helps to eliminate another important lens defect, *spherical aberration*. A lens that is unable to bring rays of light entering the edges of that lens, and those entering the centre, to a common focus exhibits spherical aberration (fig. 13). This obviously reduces image quality if left uncorrected.

ACHROMATIC OBJECTIVES

The standard type of objective used on student microscopes is the *achromat*. These have some correction of both chromatic and spherical aberration. They are designed to give their best performance with green light, which is of an intermediate wavelength between blue and red light.

APOCHROMATIC AND FLUORITE OBJECTIVES

The best research microscopes use *apochromatic* objectives which are superior to achromats. Three different colours of light are brought to a common focus. Apochromatic objectives are of complex design, being composed of many individual lenses. They are, thus, extremely expensive. A third type of objective lens, intermediate between achromats and apochromats, can still be found. These *semi-apochromatic* or *fluorite* objectives are better than achromats but not quite as good as apochromats. Such objectives used to contain at least some lens elements made of the mineral fluorite.

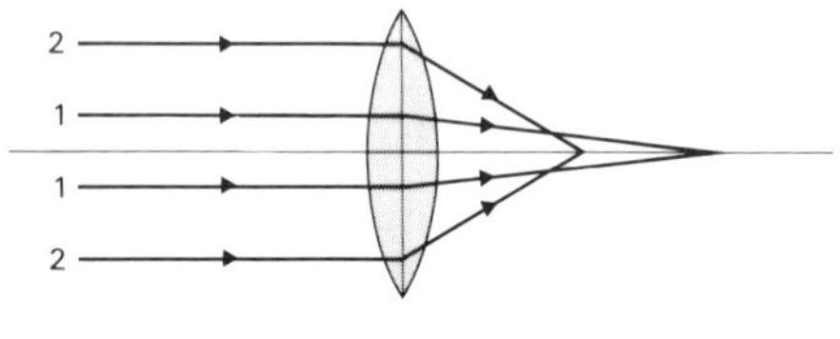

Fig. 13 Spherical aberration of a lens. The central light rays (1) are brought to a different focus from peripheral light rays (2).

FLAT FIELD OBJECTIVES

In many objective lenses, the image sharpness in the centre of the field being examined is better than that at the edges. This *curvature of field* is a common defect of objectives. It can be overcome by using flat field objectives, which generally have the word 'Plan' engraved upon them. Thus one can buy Plan-achromat and Plan-apochromat objective lenses. These are much more expensive than the achromats and apochromats described earlier.

COVERSLIP THICKNESS

Certain high magnification objectives are sensitive to the thickness of the thin layer of glass which covers the specimen (the coverslip). The required coverslip thickness is engraved on such objectives. A figure of 0·17 refers to a coverslip 0·17 mm thick. This is the thickness of the coverslip known as normal no. $1\frac{1}{2}$ type. Sometimes objectives contain movable elements to compensate for the spherical aberration produced when using a coverslip of a different thickness.

DEPTH OF FOCUS

Just like camera lenses, microscope objectives have a *depth of focus*. This is the depth of the specimen which is in focus at one time. Generally, the depth of focus is inversely proportional to the magnification, so that the

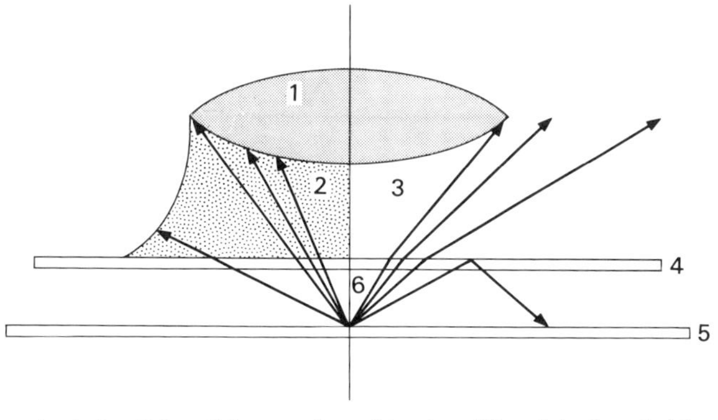

Fig. 14 The principle of the oil immersion objective. The right-hand side shows refraction of light from the specimen as the light travels from glass to air. Some of the light refracted in this way does not enter the objective. This can be avoided by interposing an oil of similar refractive index to glass, between the coverslip and the front lens of the objective, as shown on left side. (1) Objective. (2) Oil. (3) Air. (4) Coverslip. (5) Slide. (6) Specimen.

greater magnification, or numerical aperture of an objective, the smaller the depth of focus.

USING AN OIL IMMERSION OBJECTIVE

If you own a good quality microscope and want to examine small details, such as the internal component of a cell, at magnification of about 1000 times, a special oil immersion objective is required (fig. 14). These objectives have a numerical aperture greater than 1 and generally magnify from 90 times to 100 times.

Obtain a bottle of immersion oil from the manufacturer of the objective lens or buy a reputable brand from a microscope supplier. Cedarwood oil can also be used. A great deal of care is required when using an oil immersion objective. First examine the specimen in air under low magnification (for example, a × 10 objective) and select and centralise the area of interest. Rack the objective up about 10 mm (or rack down the stage if the focusing is accomplished by this means) and bring the oil immersion objective into position. Using a glass rod, place a small drop of the immersion oil on the specimen slide directly beneath the objective. With the eyes level with the specimen slide, carefully lower the objective (or raise the specimen stage) until the oil makes contact with the lens. Move the objective a little further into the oil without touching the glass specimen slide (fig. 15). Because the working distance of oil immersion objectives is very small, extreme care is now required to bring the

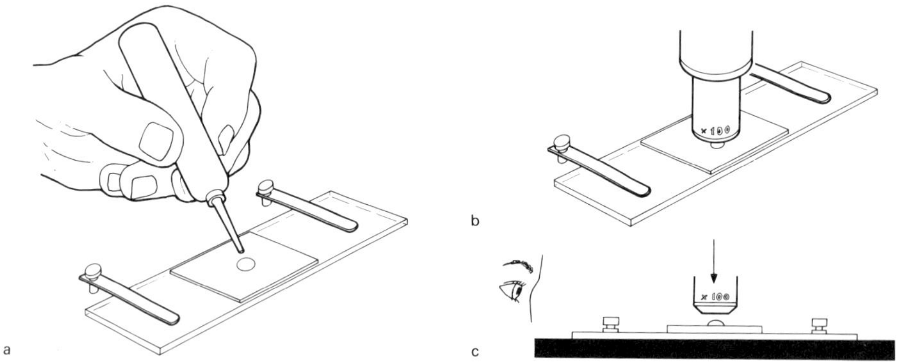

Fig. 15 The use of the oil immersion objective. (a) After viewing the slide under low power, increase the separation between the slide and the objective, and place a drop of immersion oil on the coverslip immediately above the feature of interest. (b) Swing the oil immersion objective into place over the oil droplet. (c) With your eye level with the stage carefully lower the objective into the oil droplet.

specimen into focus. If care is not taken, the specimen and objective may be damaged. Specimens are usually replaceable but the objective lens may require expensive repairs by the manufacturer. Having lowered the objective, look through the eyepiece to see if any image is visible. Using the fine focus control, move the objective upwards until the specimen is in focus. If, after raising the lens a millimetre or so, the image of the specimen still has not been seen clearly, start again. Lower the objective again into the oil and repeat the process. The amount by which the objective can be lowered will come with practice. The above procedure should never be rushed.

When you have finished viewing the specimen through the oil immersion lens, move the objective away from the slide. The front lens element of the oil immersion objective is now covered in oil and this must be removed or it may dry and form a hard film, which will affect image quality when the objective is used again. Using a lens tissue, carefully wipe the oil off the objective lens. If the oil has hardened, moisten a lens tissue with xylene and carefully wipe off. Be careful not to get oil on the other objective lenses, as this will also affect image quality. If this does occur, again carefully wipe the lens with a lens tissue.

The Eyepiece or Ocular

The eyepiece of a compound microscope enlarges the primary image formed by the objective, so that it can be recorded by the eye of the microscopist or the film of a camera. As with objectives, there are several types of eyepiece.

The *Huygenian eyepiece* is the most common type and consists of a metal cylinder with plano-convex lenses (flat on one side and convex on the other), with their convex surfaces towards the objective; they have a

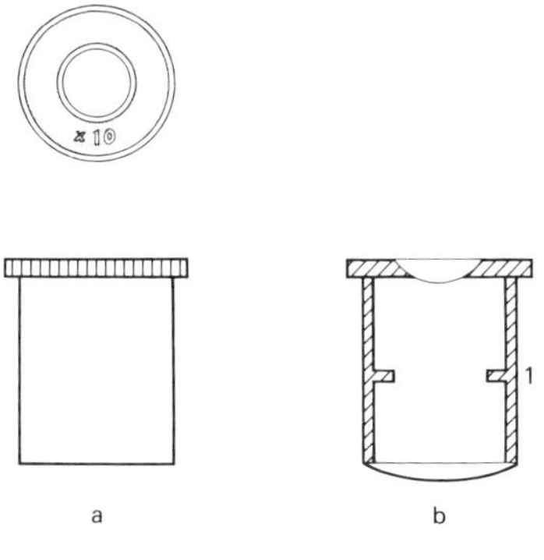

Fig. 16 (a) External features of an eyepiece. (b) Sectioned Huygenian eyepiece, containing two plano-convex lenses and a field stop (1) between them.

circular opening between them called the *field stop* (fig. 16). An engraved magnification (for example, ×10) is marked on the metal surrounding the upper lens.

Compensating eyepieces are required for apochromatic objectives, as it is not practicable to correct all the aberrations within the apochromatic objective itself. Some of the aberrations are compensated for in the eyepiece. Such eyepieces are usually engraved with the word 'compensating'.

The Condenser

A *condenser* is a lens system used to control the light before it reaches the specimen. The condenser is therefore beneath the specimen stage. Older microscopes and some modern student microscopes have a mirror below the condenser to direct light from an external source such as a desk lamp or daylight. Never use direct sunlight because this may permanently damage the eyes. Most modern microscopes have built-in illumination located in the base of the microscope stand.

Some microscopes use a translucent screen above the bulb in order to illuminate the specimen evenly. This diffuser system is adequate for general work, but for increased resolution a substage condenser system is necessary. The condenser concentrates light on to the specimen and allows the full aperture of the objective lens to be used, thus optimising the attainable resolution of fine detail (fig. 17). Condensers can be moved up

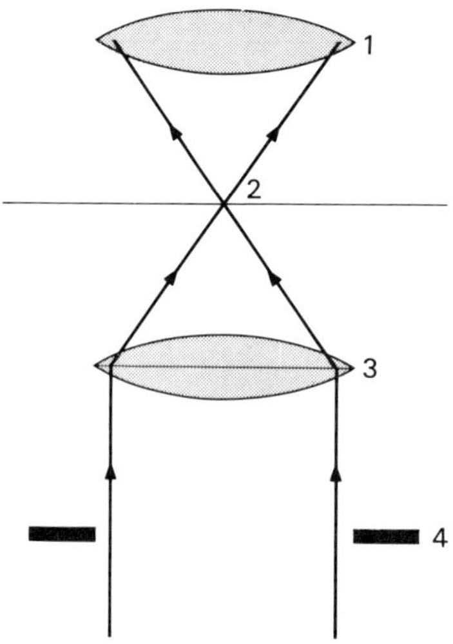

Fig. 17 The condenser (3) converges light from the lamp on to the specimen (2), and the cone of light produced fills the aperture of the objective (1). Beneath the condenser is the iris diaphragm (4).

and down beneath the specimen stage to focus the light source on to the specimen.

As with objectives and eyepieces, several different types of condenser are available. The one usually fitted to student microscopes is called an *Abbé condenser*, which is uncorrected for aberrations. The *achromatic condenser* is the best available and is corrected for both chromatic and spherical aberrations. A condenser of intermediate quality, the *aplanatic condenser*, may be found on some microscopes.

Some substage condensers are fitted with centring screws so that illumination on the specimen can be centred. Like objectives, condensers are often marked with their maximum numerical aperture. This should at least equal the numerical aperture of the objectives in use. The numerical aperture of the condenser can be adjusted by opening the *iris diaphragm* located beneath the condenser. Just as some objectives are designed for use with oil between the specimen and the lens, so some condensers are designed for use with oil between the upper condenser lens and the base of the microscope slide. For most ordinary work such condensers are tedious to use and need not concern us further.

ADJUSTING THE CONDENSER

To attain the maximum resolution from an objective, it must receive light from the specimen at an angle wide enough to fill the aperture of that lens. This criterion means that the microscopist must carefully adjust the illumination system of the microscope in order to produce the best results. The adjustment of the illumination system of a microscope depends on the type of microscope you are using. Older microscopes have a substage mirror and a condenser with iris diaphragm, whereas many modern microscopes have such additional features as a centrable lamp, lamp condenser and field diaphragm. We will therefore deal with each separately.

MICROSCOPES WITH MIRRORS

Daylight is suitable for using a microscope out of doors or near a window. The mirror usually has a plane (flat) and a concave (dished) side. The plane mirror must be used unless no condenser is fitted, in which case the concave side should be used instead. When using low magnification objectives (×4 and ×10), the concave mirror alone is adequate, but with higher magnification, a condenser is essential and this must be used with the plane mirror.

An electric light source is preferable to daylight, which is too variable. A

60 watt opal bulb in a desk lamp is excellent if the following procedure is adopted:

1 The lamp should be placed approximately 150 mm to 250 mm from the plane mirror of the microscope. The lamp should be in line with, and at the same level as the mirror.
2 Place a glass specimen slide on the stage and, using a ×10 objective about 5 mm above this, look through the eyepiece. With the condenser in a position close below the specimen, rock the mirror from side to side at about the 45° position to the vertical, until a spot of light appears in the eyepiece of the microscope. If the mirror is correctly positioned, the image through the eyepiece should be uniformly illuminated. If this is not so, gently tilt the mirror until the most even illumination is obtained. Focus the specimen as sharply as possible using the focusing controls.
3 To achieve more contrast in the specimen, remove the eyepiece and look down the microscope body tube. A disc of light should be discernible in the back lens of the objective. Close the iris diaphragm beneath the condenser until the disc of light diminishes. Carefully rack the condenser up and down until the periphery of the disc is sharply defined. If the illumination through the back lens of the objective is off-centre, then adjust the condenser centring controls, if fitted. Adjust the diaphragm until it obscures about a third of the possible field (fig. 18) and replace the eyepiece. Some further slight adjustment of the mirror and focus controls may be necessary, but the specimen should be evenly illuminated and clearly defined.

KOHLER ILLUMINATION

Modern research microscopes use a system of illumination called *Kohler illumination.* This provides the best quality of image and the highest resolution. It also provides an evenly illuminated specimen when taking

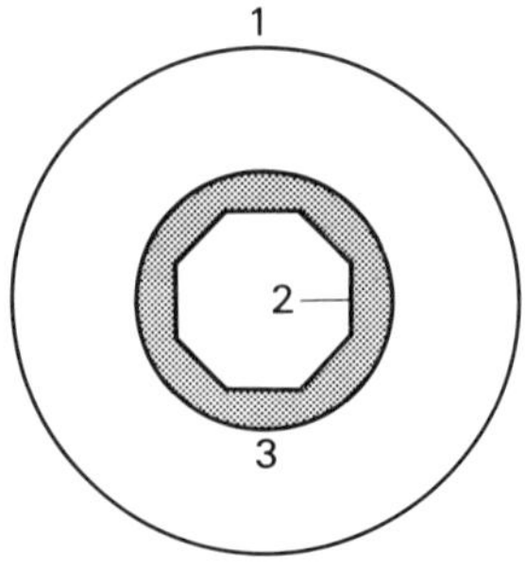

Fig. 18 The view down the body tube of a microscope with eyepiece removed, showing optimum opening of the iris diaphragm observed in the back focal plane of the objective lens. The diaphragm should obscure one third of the total diameter of the lens, as shown. (1) Body tube. (2) Iris diaphragm. (3) Objective lens.

photographs through the microscope. Many student microscopes with illumination systems in the microscope base have some variation of the Kohler type of illumination (fig. 19). Basically, this type of illumination consists of a lamp, which has a condenser lens placed in front of it (the *field* or *lamp condenser*) and an iris diaphragm (the *field diaphragm*). These are in addition to the usual substage condenser and diaphragm. The field condenser is used to focus an image of the field diaphragm on the substage condenser, which then focuses an image of this in the plane of the specimen. The lamp condenser thus acts as the source of specimen illumination.

For a microscope with this type of illumination, carefully follow the instructions given below for setting the illumination. If the microscope has a mirror and condenser, Kohler illumination can still be used, but it will be necessary to buy or construct a lamp housing with a focusable condenser and iris diaphragm. If constructing such a lamp, use a clear 100 watt bulb, and if possible, this should be capable of being centred. A diagram of this type of lamp is illustrated in fig. 19.

SETTING UP THE MISCROSCOPE FOR KOHLER ILLUMINATION

1. Place lamp, described above, about 250 mm from the microscope mirror, switch on, and adjust, until the light is centred on the mirror.
2. Adjust the mirror so that some light is directed on to the specimen.
3. Using a $\times 10$ objective, focus on the specimen.
4. Using a thin piece of white card placed as close to the substage condenser diaphragm as possible, focus the lamp condenser until a sharp image of the field diaphragm is produced on the card. Remove the white card.
5. Stop down the field iris to a small aperture. Look down the eyepiece and move the substage condenser up and down until a sharp image of the field diaphragm is visible. Centre the image of the field diaphragm by careful mirror alignment.
6. Open up the field diaphragm until its diameter is equal to, or slightly greater than the field visible through the eyepiece.

The microscope is now set up for Kohler illumination. To take full advantage of this, the aperture of the objective must be filled. Therefore, remove the eyepiece and adjust the substage condenser diaphragm control until the diaphragm obscures about one third of the field of view, as described earlier. Note that the field iris and substage condenser iris will need readjustment if the objective lens is changed for one of higher or lower magnification.

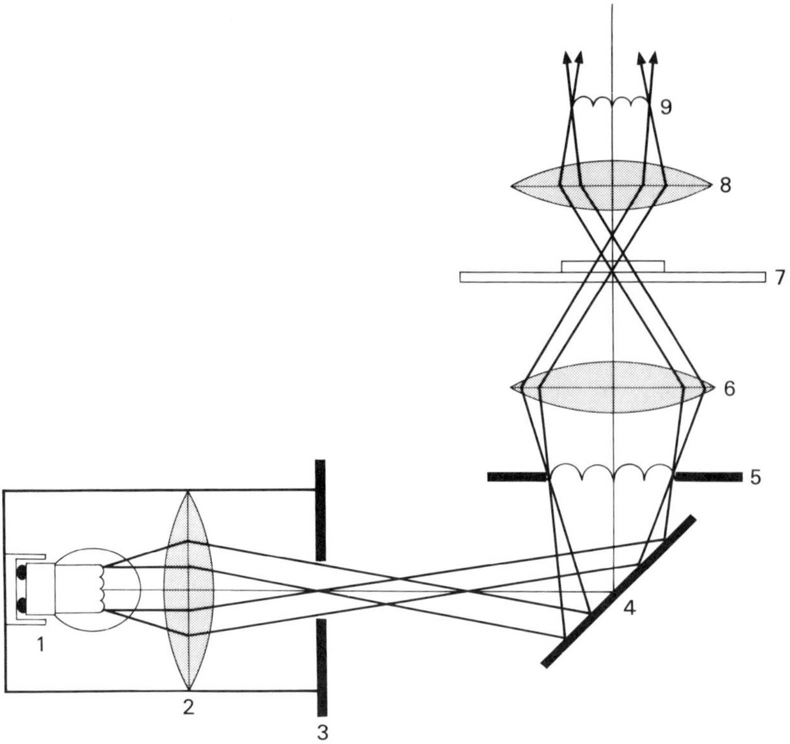

Fig. 19 The Kohler method of bright field illumination. The filament of the lamp (1) is focused at the iris diaphragm (5) which is at the front focal plane of the condenser (6). The filament image is projected on to the back focal plane of the objective (9) by the condenser and the objective (8). This method of bright field illumination is preferable for successful photomicrography. (1) Lamp. (2) Field condenser. (3) Field diaphragm. (4) Mirror. (5) Iris diaphragm. (6) Condenser. (7) Specimen. (8) Objective. (9) Image of filament in back focal plane of objective.

Resolution Limit of the Compound Microscope and Choice of Lenses

Beyond a certain limit there is no point in further magnifying an object. This is because no more information or fine detail can be seen. Magnifying an object beyond the point where no further resolution is available is called *empty magnification*. Consider a photograph in a newspaper which is composed of dots of varying density. When viewed normally, the photograph conveys information, but look at the photograph again using a magnifying glass. Now the dots are visible but it is not possible to extract any more picture detail.

A good quality light microscope, correctly adjusted, can resolve detail in an object if it is greater than a quarter of a micrometre in size. Obviously there is no point in using magnifications greater than that required to make this separation of object detail visible to the eye. Thus a useful rule is not to let the total magnification exceed 1000 times the NA of the objective in use. For example, using a ×40 objective with an NA of 0·65 (1000×0·65 = 650), then a ×10 eyepiece (10×40 = 400) would be more appropriate than a ×20 eyepiece (20×40 = 800).

In practice, field of view, working distance and depth of focus all play a part in the choice of lenses, which with experience becomes automatic.

Measuring Objects under the Compound Microscope

It is often necessary to measure the size of microscopic objects. This can be achieved by measuring the size of an object in a photomicrograph and dividing this by the total magnification of the image of that object (see Chapter 7). If a measurement is required more quickly and accurately, use an *eyepiece micrometer graticule*, which has been calibrated using a *micrometer slide*. These items are available from some of the suppliers listed. The eyepiece graticule is marked, but these marks have no particular measurement value (fig. 20). The eyepiece graticule has to be placed inside the eyepiece so that it is in focus at the same time as the specimen. The eyepiece usually consists of two lenses which have a fixed diaphragm (the field stop) between them. Unscrew top lens and carefully drop eyepiece graticule down the inside of eyepiece tube so it rests on the field stop. Replace top lens of the eyepiece. Look through eyepiece to check that the scale is in focus to your eye. If not, the field stop can sometimes be carefully pushed up and down the eyepiece tube until the desired focus of the graticule is accomplished. Otherwise a focusing eyepiece is required.

The micrometer slide has a scale on it divided into 0·01 or 0·1 mm units. The micrometer slide is placed on the specimen stage and brought into

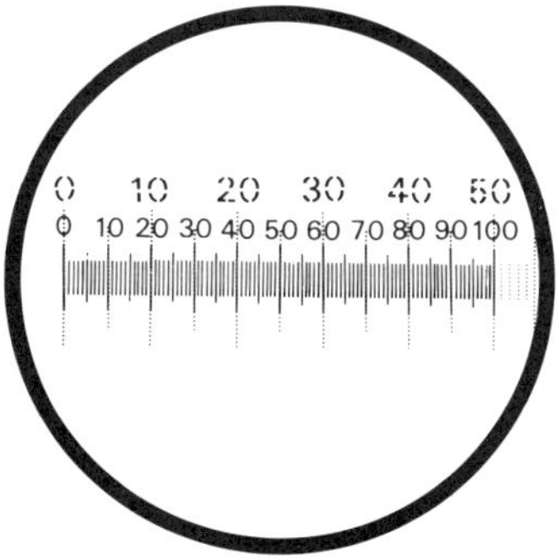

Fig. 20 A micrometer slide (large stippled figures) being viewed through an eyepiece fitted with a micrometer graticule (small solid figures).

sharp focus. The eyepiece scale, or divisions on this scale, can now be compared to the known distances marked on the micrometer slide (fig. 20). Thus if a specimen is examined at the same magnification as the micrometer slide, the size of features on the specimen can easily be calculated. The scale in the eyepiece must be calculated again for each of the objectives of the microscope.

MAKING STAGE MICROMETERS

Stage micrometers are relatively expensive items. Instead of purchasing one it is possible to make a micrometer photographically.

Photograph a 0–100 mm scale on a transparent ruler. The negative produced can be further reduced in a photographic enlarger and printed on another piece of negative film material (thus producing a transparent positive) so that the 0–100 mm scale is reduced to 10 mm. This 'positive' can be mounted on a glass microscope slide beneath a coverslip with Canada balsam and used as a stage micrometer.

Care of the Microscope

A microscope is a delicate and complex instrument that must be treated with care if you wish to obtain consistently good results.

1 After use, it should be covered in a plastic bag or replaced in a box or cupboard to keep it dust free. Grit or dust can damage mechanical mechanisms and scratch the glass lenses.
2 Dust can be removed by using a blower brush which is easily obtained from a photographic shop. Cleaning the lenses and mirror should be done with great care, and only with the blower brush and lens cleaning tissues. After blowing the dust off a lens surface, wipe (never scrub) with a clean piece of lens tissue.
3 Immersion oil should be removed from objectives (see earlier section on *Using an Oil Immersion Objective*).
4 Try to avoid putting greasy fingerprints on the lens or mirror surfaces.
5 The mechanical controls of a microscope may occasionally require cleaning and lubricating. If this becomes necessary, use only a minimum amount of lubricant.
6 The microscope should always be lifted or carried by the stand, with one hand beneath the foot or base for additional support. Avoid picking up the microscope by its movable body tube, this could damage the focusing mechanism.
7 To avoid dust and damage, lenses and accessories should be stored in individual boxes or a box which has several compartments.
8 If water is spilt on stage, wipe it off immediately to avoid corrosion.

2
Constructing a Home-made Microscope

Quality at Low Cost

For those who cannot afford a good quality microscope or who have been unable to find a good second hand microscope, there is a third alternative. It is possible to construct a microscope of good optical quality with only moderate manual skills. The quality of the objective is the most important part of any compound microscope, but many of the cheap microscopes on sale have objective lenses possessing unacceptable chromatic and spherical aberrations, and such poor quality can easily dampen enthusiasm for microscopy. Money would be better spent in buying a couple of objectives and a Huygenian eyepiece from one of the microscope suppliers listed in Appendix II, or from colleges or industry if they are disposing of old equipment. These would then form the basis of a home-made microscope. The maximum usable magnification of such a microscope is ×400, so appropriate lenses to purchase would be a Huygenian eyepiece of ×10 magnification, and ×10, ×20 and ×40 achromat objectives. These items can be bought singly or all together, depending upon available finances. Once suitable eyepieces and objectives have been purchased the microscope can be easily and cheaply constructed, and a great deal of satisfaction can be obtained from building it.

This do-it-yourself microscope is capable of producing excellent service and is easily repaired or modified. Its optical performance depends on the quality of construction and the quality of the lenses purchased. It also makes an ideal microscope for photography through the microscope as described in Chapter 7.

Making the Body Tube

This is made from a length of domestic copper pipe approximately 28 mm ($1\frac{1}{10}$ in.) external diameter. The mechanical or body tube length of many microscopes is 160 mm and is measured from the base of the screwthread of the objective to the lower lip of the eyepiece lens. Cut the copper pipe to precisely 160 mm. Some microscope objectives are designed for use with a tube length of 170 mm, so check designated length when buying lenses.

Copper piping varies in its wall thickness, so the internal diameter may allow the eyepiece to fit without modification. If this is not the case, because the internal diameter is too great, a suitable diameter piece of wood (approximately 25 mm (1 in.) long) can be glued into the top of the tube. Araldite or similar epoxy resin is ideal for this purpose. When dry, a 23 mm ($\frac{9}{10}$ in.) diameter hole for the eyepiece should be bored through the centre of the wooden insert using a brace and bit. The eyepiece should slide into place without being too loose, and its sides must be parallel to the copper tube.

Inserting the objective is a little more difficult. The objective lens has a screw fitting approximately 20 mm ($\frac{4}{5}$ in.) in diameter. The easiest way to fit it into the tube is by (again) gluing a suitable piece of wood into the tube and when dry, boring a central hole parallel to the sides of the tube so that

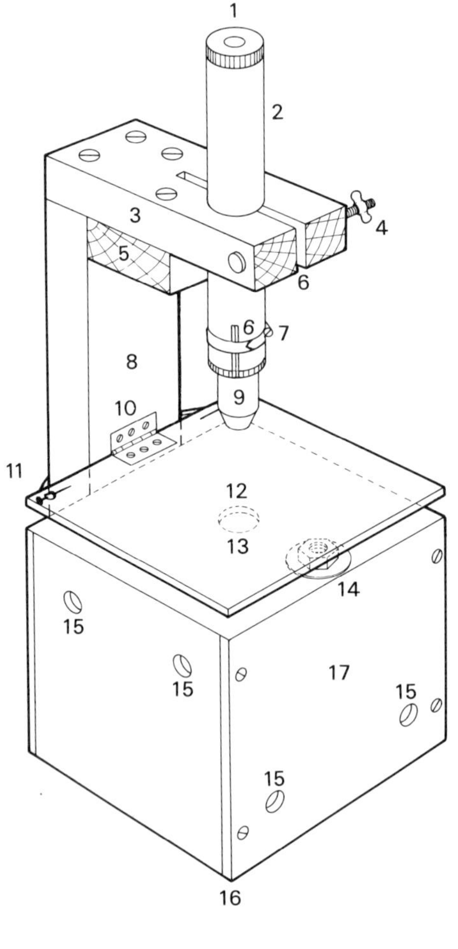

Fig. 21 The major parts of the home-made microscope described in the text. (1) Eye-piece. (2) Body tube. (3) Horizontal beam. (4) Bolt and wing nut. (5) Supporting block. (6) Saw-cut slit. (7) Objective lens retaining clip. (8) Upright stand. (9) Objective lens. (10) Specimen stage hinge. (11) Rubber band. (12) Specimen stage. (13) Illumination hole in illumination box. (14) Fine focusing mechanism. (15) Ventilation holes. (16) Illumination box. (17) Detachable front of illumination box.

the threads of the objective can be pushed into the hole, while further insertion is stopped by the shoulder surrounding the lens. Now cut a slot approximately 20 mm long through the centre of the tube, parallel to its long axis (fig. 21). Different objectives can now be secured in the tube by using a suitably sized, adjustable clip (jubilee or hose-clip from a car accessory shop) over the slot to constrict the tube. Tightening the clip closes the slot in the tube, so that the wooden insert just grips the threads of the objective lens and holds it securely in place. This method of fixing the objective has a number of advantages, in that the lens can easily be changed and misalignment of the lens can be corrected before tightening the retaining clip.

Making the Stand and Illumination System

The illumination system comprises a lamp holder and a small 15 watt opal lamp contained in a box; the size of box depending on the size of lamp and holder. The box is constructed from 9 mm ($\frac{3}{8}$ in.) plywood and should be about 125 mm (5 in.) high and 125 mm (5 in.) square. Ideally the box should be glued and nailed together. The front of the box (that part opposite the upright stand) should be detachable for centring the illumination and for changing the lamp. Ventilation holes must be drilled into this box to allow heat dissipation when the lamp is on (fig. 21). It is desirable to line the inside surface of the top of the box with metal foil to reflect the heat. A hole of approximately 12 mm ($\frac{1}{2}$ in.) diameter should be drilled through the centre of the top of the box and the light fitting centred below this hole. This is important in order to achieve even illumination of the specimen.

The upright stand which supports the body tube must be substantial, and accurately made. The stand should be constructed of hardwood not less than 62 mm × 25 mm ($2\frac{1}{2}$ in. × 1 in.) in section. The upright should run from the base of the box (approximately 125 mm (5 in.)) for secure attachment, and protrude 140 mm ($5\frac{1}{2}$ in.) above it. A horizontal beam 125 mm (5 in.) long is attached at right angles to the top of this upright. This must have a hole the diameter of the body tube drilled into it, approximately 50 mm (2 in.) from one end, in which the body tube fits. A vertical slit is cut through the centre of the horizontal beam from the front to about 12 mm ($\frac{1}{2}$ in.) beyond the body tube hole. A bolt hole is drilled at right angles through the front beam, through the vertical slit. A suitable bolt is pushed through this hole and a washer and wing nut screwed on to the protruding end. When the body tube is pushed through its hole, the wing nut can be tightened to secure the tube in place, but can be slackened to allow for vertical movement to accomplish coarse focusing. A second adjustable clip can be attached around the top of the body tube, to act as a stop in case the tube drops towards the specimen slide. The horizontal beam must be accurately fitted to the upright stand, so that the objective is centred symmetrically on the hole in the top of the illumination box, so that it is evenly lit. The horizontal beam should be glued and screwed into position. A supporting block must be attached into the right angle formed between the vertical stand and the horizontal beam.

Making the Fine Focusing Mechanism

Rough focusing is accomplished by raising or lowering the body tube, then clamping it with the wing nut on the horizontal beam.

Fine focusing is essential, and this was our most difficult design problem as it was necessary to have a mechanism which was within the ability of someone without extensive workshop facilities. The most precisely machined, readily available component is a nut and bolt, and this is the basis of the fine focusing mechanism. Use a bolt with a fine thread pattern (a large number of threads per unit length, eg BSF threads) approximately 30 mm long × 7 mm diameter. A central hole is drilled through the front edge of the top of the illumination box about 100 mm (4 in.) from the rear upright. The bolt is pushed through this hole from below and firmly attached by a nut. A further nut is required, to which is glued a large washer. The kind of body washer used to attach rear seat belts in cars is ideal, and these can be obtained from a car accessory shop. A suitable glue is cyanoacrylate ester (Loctite 'super glue – 3') which bonds in seconds. This particular nut is loosely screwed on to the protruding bolt with the large washer on the undersurface of the nut.

The specimen stage-plate comprises a piece of 3 mm acrylic sheet (ICI 'perspex') approximately 112 mm × 100 mm (4½ in. × 4 in.). The size of this acrylic sheet should be measured from the front of the upright stand to the front of the bolt on the illumination box. In the microscope's simplest form, this transparent acrylic sheet should be made translucent by scrubbing one surface with household scouring powder. The translucent sheet acts as a diffuser to spread the light from below and thus evenly illuminates the specimen. The acrylic sheet is attached by a suitably sized hinge to the upright stand, in such a way that the sheet is parallel to the top of the illumination box and about 3 mm (⅛ in.) above the top of the fine focusing bolt. The nut with the large washer should be unscrewed until the nut protrudes above the end of the bolt. By rotating the large washer the stage plate should move up and down.

It was found necessary to attach a rubber band around the upright stand, through the rear corners of the acrylic sheet, to minimise the play in

Fig. 22 The simpler version of the home-made microscope in completed form. For identification of parts, see fig. 21. An additional clip has been placed at the top of the tube to prevent this from falling on to the specimen stage.

the hinge. If this is not used, the image can shift too far while being viewed under high magnification.

Grease or petroleum jelly should be smeared on the fine focusing bolt. This acts as a lubricant and also takes up the play between the focusing nut and the threads of the bolt. This is necessary when using the microscope at high magnification.

The microscope should be set up with the stage capable of up and down movement (the stage will not move downwards if the nut with the large washer is screwed down below the end of the bolt). The specimen on a glass microscope slide is placed on the specimen stage-plate, and the specimen roughly focused, by slackening off the wing nut on the horizontal beam and moving the body tube carefully up and down until the image comes into focus. The wing nut is then locked and fine focusing can be accomplished by slight rotation of the fine focusing nut, using the large washer.

The basic microscope described above (fig. 22) makes an ideal instrument for beginning microscopy. A mechanical stage can be added to it to allow more precise specimen movements; and a rudimentary condenser system, with iris diaphragm, can be added to increase resolution and alter the illumination intensity.

Making the Mechanical Stage

The mechanical specimen stage is constructed from acrylic sheet, cyanoacrylate ester glue, two fine threaded bolts (75 mm long, 2BA thread size) and four nuts (2BA). The dimensions of the parts shown in fig. 23 are only for use with the 112 mm × 100 mm acrylic specimen stage-plate and with the use of 76 mm × 25 mm (3 in. × 1 in.) glass microscope slides. Mark out the parts shown in fig. 23 on acrylic sheet, and accurately cut them out, using a coping saw fitted with a blade suitable for woodwork. Accuracy in both cutting out and assembling these pieces is essential, in order to produce a mechanical stage which works smoothly and precisely without 'backlash'. Before final assembly, all parts should be assembled

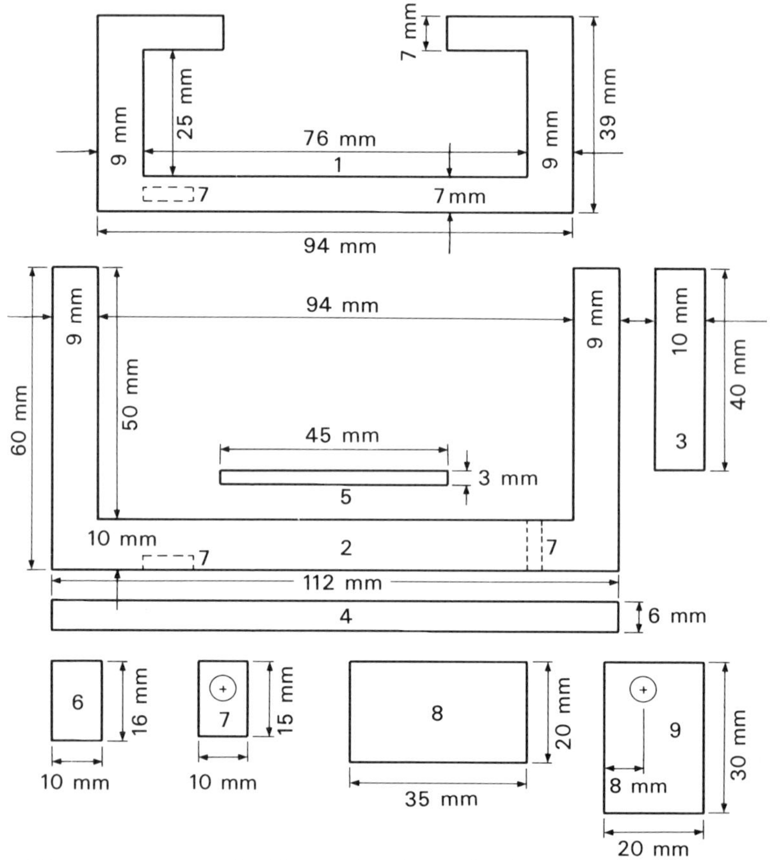

Fig. 23 These parts are required for making the mechanical stage. Part (3) two required, part (6) two required, part (7) three required. Part (7) has a 5 mm hole centred 9 mm from base. Part (9) has a 5 mm hole centred 18 mm from base, 8 mm from edge. The other part numbers are referred to in the text of Chapter 2.

without glue, so that any potential mistakes can be avoided (fig. 24). Incorrect assembly can usually be corrected by placing a wood chisel between the incorrectly assembled parts and tapping with a mallet. The parts can be reassembled after removing the old glue with a file.

ASSEMBLY OF THE MECHANICAL STAGE

1 Glue part 4 to front edge of specimen stage-plate, flush with bottom surface, so that a lip is formed above the surface of the plate (fig. 24).
2 Place (do not glue) the front of part 2 against the lip formed by part 4, and then glue part 5 into position, so that part 2 is free to slide from side

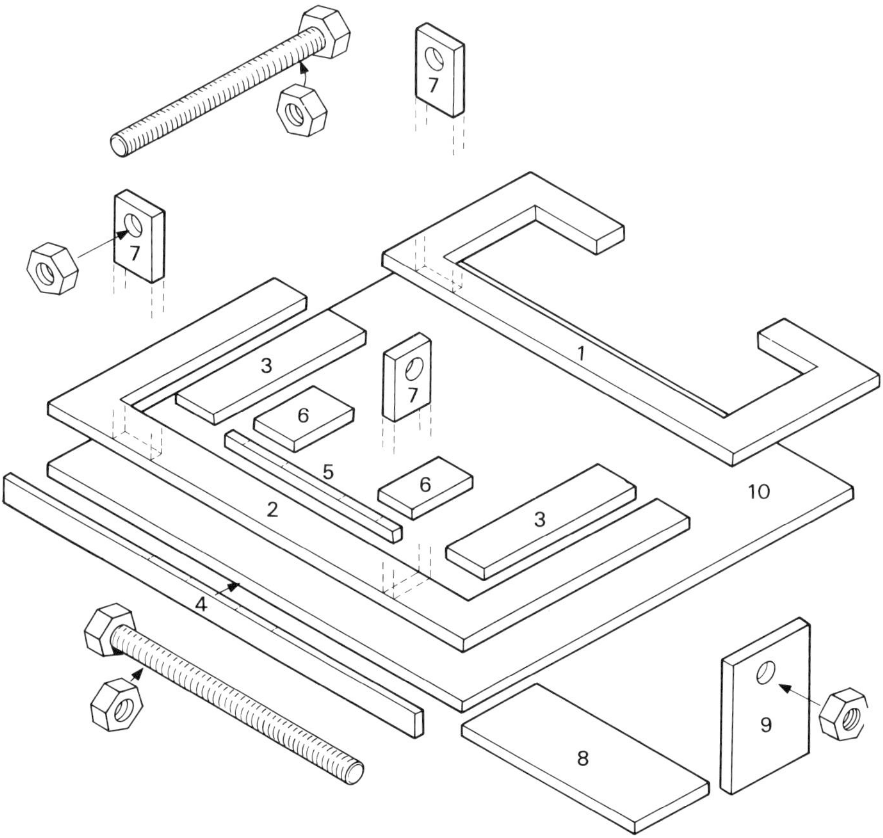

Fig. 24 Diagram of exploded view of the parts used in the construction of the mechanical stage. The stage-plate is part 10.

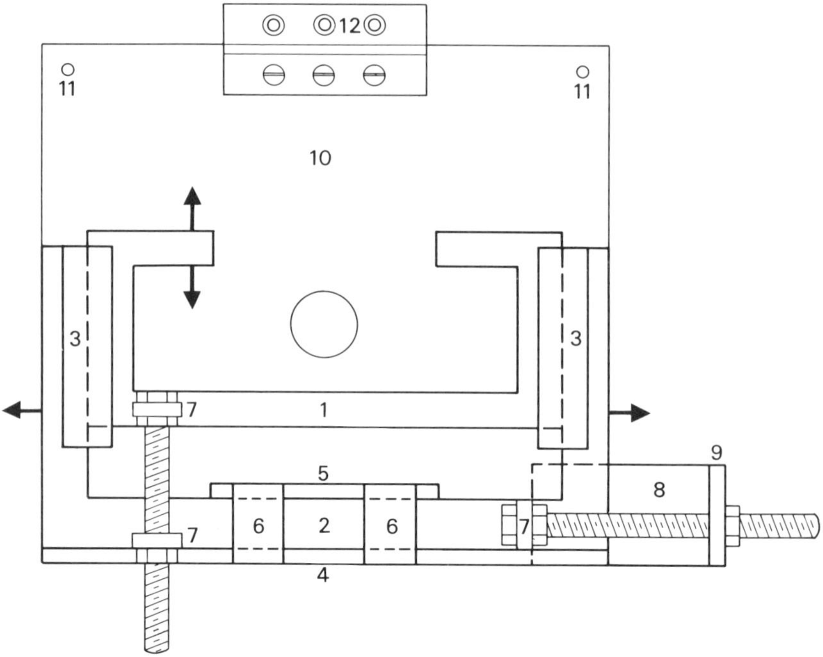

Fig. 25 Plan view of the assembled mechanical stage showing the relative positions of the numbered parts and the axes of movement. (11) shows the holes at the rear corners for the rubber band. (12) is the hinge.

to side. Part 5 must be positioned centrally, leaving 30 mm on each side, or insufficient lateral movement of the glass specimen slide will result (fig. 25).

3 Glue the retaining plates (parts 6) into position, on parts 4 and 5 only. Glue must not come into contact with part 2.
4 Place the slide carrier (part 1) into position within part 2, and glue the retaining strips (parts 3) into position on part 2, so that they overlap part 1. Part 1 must be free to move relative to part 2.
5 Glue part 8 beneath the front right-hand corner of the specimen stage-plate so that 20 mm protrudes to the right (fig. 25).
6 Glue one of the parts numbered 7, vertically to the right-hand side of part 2 in the position shown (fig. 25).
7 Glue part 9 vertically to the end of part 8 so that the hole in part 9 is at the same level as the hole in part 7 of paragraph 6.
8 Push a long bolt through the hole in part 7 and attach a nut. Now push the bolt through the hole in part 9 and attach a further nut. This second nut should be glued to part 9 using cyanoacrylate ester glue. The first

nut should be glued to the bolt near to part 7 so that they both rotate in part 7 but cannot be unscrewed relative to one another. This is best accomplished by tightening up the nut so that the bolt is free to move in part 7 without play. Mark corresponding positions on the bolt head and nut, using a pencil or a felt-tipped marker pen. Unscrew the nut and put a drop of cyanoacrylate ester glue on the threads of the bolt and screw up the nut once more until the marks correspond. After a few seconds the nut will be glued firmly in place. By rotating the threaded end of the bolt, the mechanical stage should move easily both left and right across the specimen stage-plate.

9 The two remaining parts 7 should be glued to the left side of parts 1 and 2 in the positions shown (fig. 25). The remaining long bolt should be pushed through the hole in part 7, attached to part 1, and a nut glued into position on the bolt as in paragraph 8. The bolt should then be pushed through the part 7, on part 2, and a further nut added which should be glued to the part 7, attached to part 2. The parts numbered 7 can be replaced with small metal brackets to increase strength and durability (fig. 26).

The assembled mechanical stage (fig. 26) should be capable of controlled movement, front to rear and left to right of the specimen stage-plate. The specimen stage-plate, with mechanical stage, can now be hinged to the upright stand of the microscope. If the movement of the mechanical stage is not smooth, petroleum jelly can be used as a lubricant.

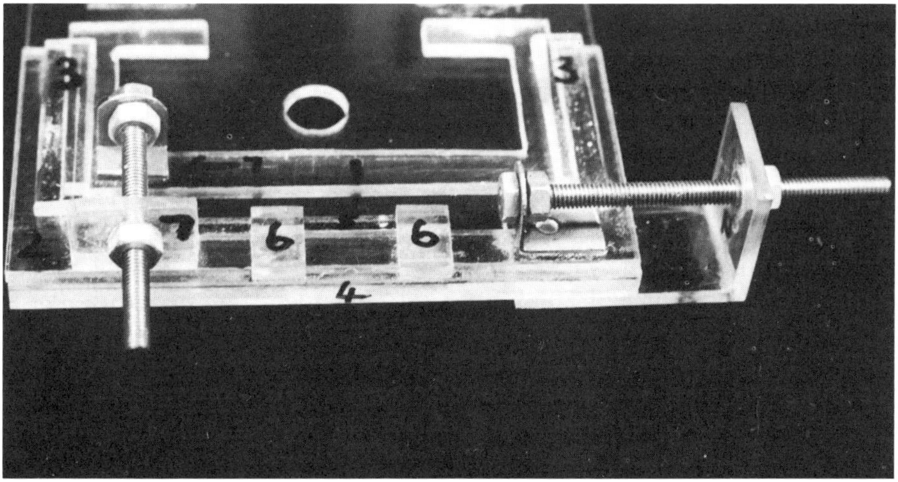

Fig. 26 The assembled mechanical stage for the home-made microscope. The numbered components can be identified from fig. 23. Note the hole in the centre of the stage plate for use with a rudimentary condenser, and the replacement of parts 7 with metal brackets.

Fig. 27 Top view of the illumination box of the more complex version of the home-made microscope, showing iris diaphragm, condenser lens, and fine focus washer with attached nut.

Adding a Condenser and Iris Diaphragm

To achieve maximum resolution, a condenser lens system can be added to the microscope. All that is needed is a suitable convex (converging) lens of approximately 25 mm (1 in.) focal length. These can sometimes be found by dismantling old redundant camera lenses. A hole of approximately 12 mm ($\frac{1}{2}$ in.) should be drilled through the acrylic specimen stage-plate beneath the objective lens (fig. 26). The convex lens is placed on top of the illumination box, so that light transmitted from the lamp is converged on to the specimen, through the hole in the specimen stage-plate (fig. 27).

An iris diaphragm can be added to the condenser to control the amount of light reaching the specimen. If the condenser lens was removed from an old camera, then the iris diaphragm from this can be used (fig. 27). If this is not possible, then a strip of wood or aluminium can be used, with a series of progressively smaller holes drilled in it. This can be moved across the hole in the illumination box, to reduce the intensity of the illumination.

3
Mounting Specimens for the Microscope

Acquiring a microscope is only the first stage in microscopy. Items for examination under the microscope need to be mounted in a manner suitable for this instrument, and in a form which maximises the amount of detail that can be seen in the specimen. Preliminary treatment of specimens can vary from the relatively straightforward to the relatively complex, depending largely on the nature of the specimen and the level of detail required. It is, of course, also possible to purchase prepared slides from the suppliers listed in Appendix II and this may be useful for specimens which are particularly difficult to obtain or prepare.

The basic requirements for mounting specimens are glass slides, coverslips and a mounted needle. These are essential. The glass slides, upon which the specimens are placed, are of standard size 76 mm×25 mm (3 in.×1 in.). Coverslips, however, are of variable size and shape (round, square or oblong), and are made from extremely thin glass (see Chapter 1). The coverslip fulfils two major functions. Firstly it helps to flatten the specimen, thus ensuring that most of the specimen is in focus, when the depth of focus of the objective is small. Secondly it protects the objective from possible immersion in the fluid in which the specimen is mounted. The mounted needle is used for arranging the specimen in its mounting fluid, and also for lowering the coverslip (see below). Other instruments, whilst not strictly essential, may greatly facilitate the whole process of specimen preparation and mounting. For example, a pair of finely pointed forceps is useful not only for picking up specimens, but also for handling slides so as to avoid covering them with fingerprints. In addition, an artist's paint brush is helpful when picking up small delicate specimens which may otherwise be damaged.

Simple Dry Mounts

Although most specimens must be mounted in fluid, some may nevertheless be placed dry on to the slide, without a coverslip, and then viewed directly under the microscope without further treatment. For example, a strand of hair may be fixed to a slide by means of a piece of self-adhesive

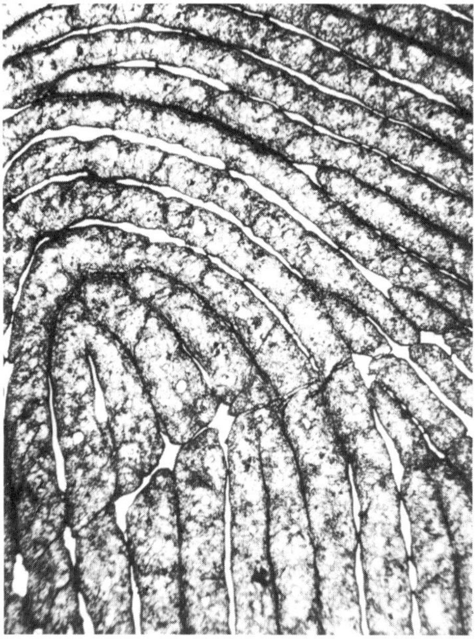

Fig. 28 Part of a human thumbprint on an acetate peel (×10).

clear tape at either end, and examined in this form. Other types of fibre may be treated in the same way.

ACETATE PEELS

The surface features of specimens can be examined easily by replicating them on acetate sheets (Cellophane). This can be accomplished by wetting the surface of interest with acetone and firmly applying the acetate sheet. After a few minutes peel the sheet from the surface, trim to size and place on a slide for examination. The peel can also be mounted in a standard transparency holder for projection (see Chapter 1). Suitable subjects for this technique include fingerprints, or any other object with a textured surface, such as unvarnished wood, leaves, cloth and so on (fig. 28), providing they are free from moisture.

Temporary Mounts in Water

Most specimens, particularly those of biological origin, cannot be mounted without some supporting medium to maintain their shape. The most common medium is water, which can be used for rapid preparation of slides, or for viewing small aquatic animals and plants (see Chapter 5). Water mounted specimens are, however, only temporary (several minutes, unless continually topped up with water) and more complex preservation procedures are necessary for making permanent mounts (see below).

In preparing a water mount, a small drop of water should be placed in the centre of a clean glass slide, using either a glass rod or a small dropping pipette. The specimen can then be placed in the water drop, and a coverslip carefully lowered on to it. Dropping the coverslip on to the specimen, or even gently lowering it parallel to the slide surface, should be avoided, as this will inevitably trap numerous air bubbles under the coverslip, thus impeding examination of the specimen. A better result is obtained by lowering the coverslip as if it were hinged to the slide on one edge, by placing a mounted needle under the free edge (fig. 29). If too large a water droplet has been used, then the coverslip will float and the

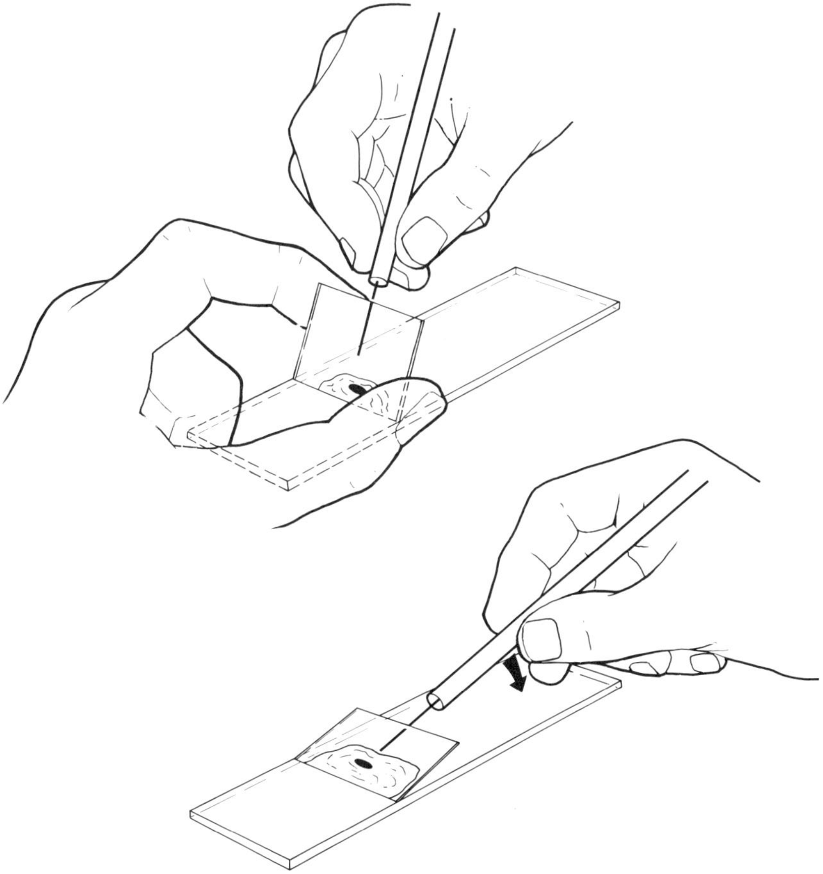

Fig. 29 Using a mounted needle to lower a coverslip.

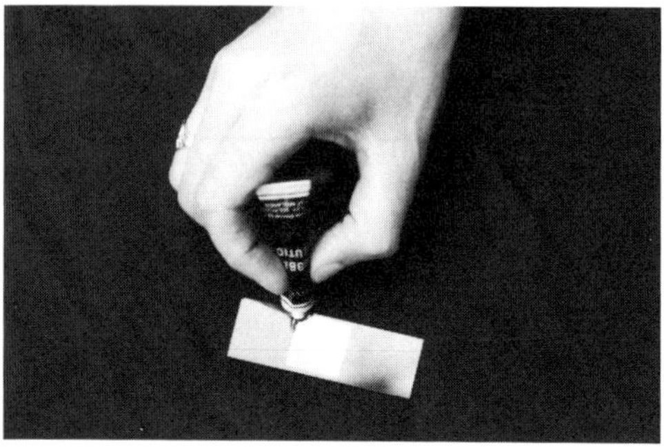

Fig. 30 Sealing a coverslip by ringing with rubber solution.

correct focusing of the specimen will be difficult, not only because of the depth of water, but also because even slight movement will jar the floating coverslip. Excess water can be removed from such a mount by touching the edge of the coverslip with absorbent paper.

An alternative problem is the drying out of the preparation as water evaporates from under the coverslip. The problem may be more acute if the microscope has a built-in lamp, which may have a substantial heating effect on the slide. Water may be added to the preparation simply by placing a droplet at the edge of the coverslip; the water will be pulled under the coverslip by capillarity. Otherwise, especially if non-living specimens are used (eg dead animals, or animal and plant parts), evaporation can be slowed by using a mixture of equal parts of glycerine and water (50% glycerine), rather than just water. Such a preparation may last several days. Water loss can also be reduced by sealing the edge of the coverslip with nail varnish applied with a brush, or with rubber solution squeezed from a tube (fig. 30).

A semi-permanent preparation (9–10 months) may be made by mounting the specimen in glycerine jelly (see Appendix 1), for which the slide and coverslip should be pre-warmed. Specimens may be transferred to glycerine jelly directly from water or from the glycerine solution without further treatment. Sealing the coverslip will render the preparation fairly permanent, although not as permanent as the resinous mounts mentioned below.

Permanent Mounts

A specimen will last indefinitely if it is thoroughly dehydrated, and mounted in a non-aqueous medium. Various resins, both natural and synthetic, are now available for making such permanent preparations. Traditionally the natural resin Canada balsam has been most widely used, and this is still a popular mounting medium.

Items to be mounted in Canada balsam and other resins must first be dehydrated, since the presence of water will lead to deterioration of the specimen. Dehydration is accomplished by passing the specimen through a series of alcohol/water mixtures in which the proportion of alcohol is

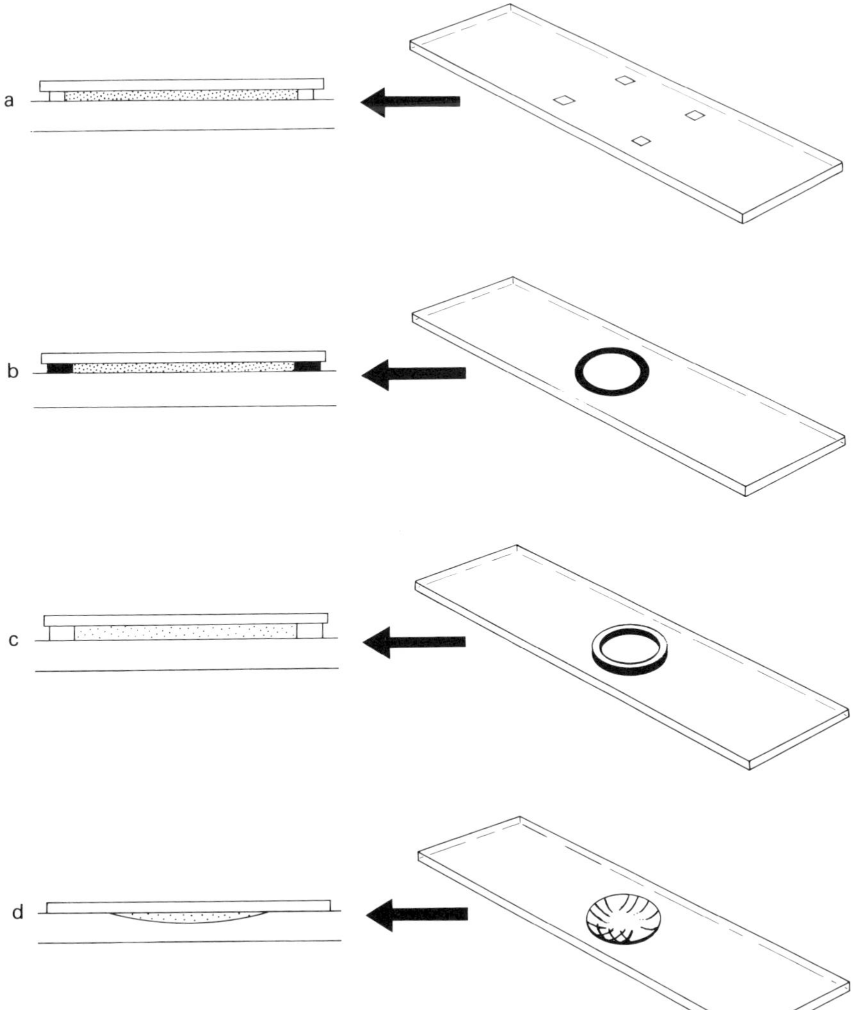

Fig. 31 Various methods used to mount thick specimens.
(a) Four fragments of broken coverslip. (b) A circle of ringing cement applied from a turn table. (c) A cemented-down metal ring. (d) A cavity slide.

progressively increased (an ascending series of alcohols). Ideally the dilutions should be made by adding the correct proportion of water to a known quantity of pure ethanol (absolute alcohol), but if ethanol is not available pure acetone can be used as a substitute. In view of the highly inflammable nature of these chemicals, careful storage and supervision during usage is essential.

The procedure for dehydration is to place specimen into 50% alcohol (or 50% acetone) for 5 minutes, followed successively by the same immersion time in 70% and 90% alcohol, and twice in 100% alcohol. The time spent in each alcohol should be increased for large specimens. From absolute alcohol or acetone, the specimen should be placed in a clearing agent such as xylene (xylol), clove oil, cedarwood oil or paraffin (pink variety) for a few minutes. Once cleared, the specimen can be transferred to a drop of Canada balsam (or other resinous medium) on a microscope slide and a coverslip applied. Some modern mounting media, for example Euparal, do not require clearing agents, so specimens may be transferred to them directly from absolute alcohol. The mountant must now be allowed to harden, a process which takes several months at room temperature, but can be achieved in a matter of hours in an oven at 60 °C. Examples of the use of this technique will be given in Chapter 4.

MOUNTING THICKER SPECIMENS

Some specimens may be too thick to allow a coverslip to be placed flat upon them. This can be overcome by supporting the coverslip at the required height by means of some other material. If the necessary height increase is very small it can be achieved by breaking a coverslip, and placing a small fragment at each corner where a coverslip is to be lowered over the specimen. For a higher mounting, a ring of plastic or some other impermeable material (for instance, a washer) can be used to support the coverslip (fig. 31). Cardboard rings are unsuitable for this, as they are permeable, and any water which is allowed into the preparation will eventually lead to its deterioration.

Alternatively, a cavity slide may be purchased for this sort of preparation. The cavity in the slide allows for the mounting of a fairly large specimen without the need to raise the height of the coverslip (fig. 31).

The Use of Stains

For most biological materials, magnification alone is not sufficient to allow the discrimination of fine structural details, since colour and contrast are often lacking. Cells in particular usually appear virtually trans-

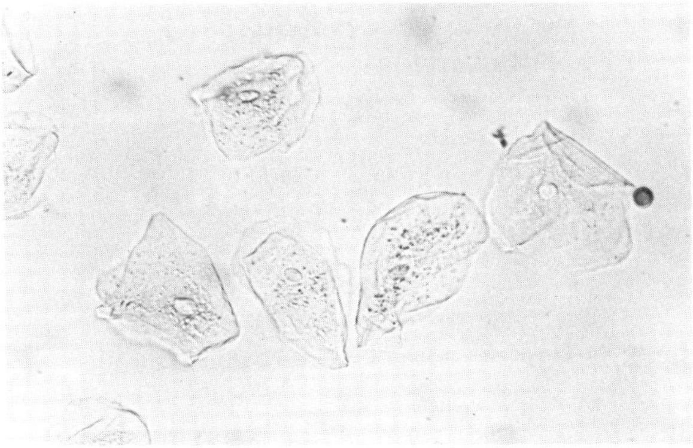

Fig. 32 Unstained cheek epithelial cells in normal transmitted light (×200).

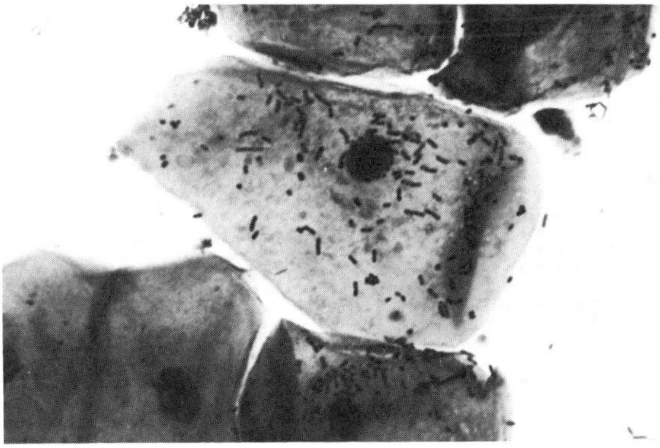

Fig. 33 Cheek epithelial cells stained with methylene blue, showing the prominent nuclei. Compare with the unstained cells in fig. 32. Notice that the surface bacteria are also densely stained (×500).

parent, and even such major structures as nuclei are visible only with difficulty (fig. 32). For this reason it is necessary to stain specimens, not only so they can be more easily seen, but also to enhance the contrast between different areas of the specimen (fig. 33). Many stains are selective, being preferentially taken up by specific parts of the specimen, others are more general and stain all parts. In the latter case, contrast is still improved since the stain aggregates in the denser structures, which thus appear darker.

Some stains have been developed specifically for use with living tissues or organisms. These are called *vital stains* and, since they are generally water-based, they are unsuitable for staining specimens destined for mounting in a resinous medium.

An enormous variety of stains is now available, and few generalisations about their use can be made. Instead, examples of specific staining techniques will be given in this and subsequent chapters. The examples have been chosen because of their widespread use and because the stains can be easily obtained, either ready-made from the suppliers listed in Appendix II, or else by making them up from their constituents (Appendix I).

Some Examples of Stained Mounts

HUMAN CHEEK CELLS

The inside lining of the cheek is a good source of loose epithelial cells, which may be obtained by gently scraping a finger nail or a sterile blunt object, such as a glass rod, across the cheek lining. The cells thus obtained may be spread on the surface of a slide. Loose cells in the fluid are visible without staining, but the contrast between the nucleus and the cytoplasm of the cells can be enhanced by staining with methylene blue. A couple of drops of methylene blue should be added to the cells and left for about a minute, after which the excess stain may be removed by tilting the slide

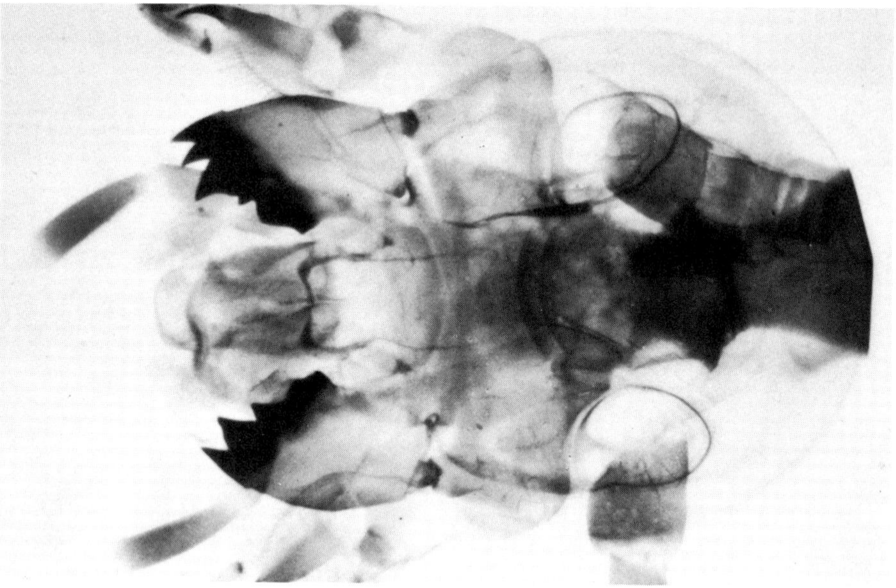

Fig. 34 Unstained whole mount of the head of the cockroach *Periplaneta americana* after boiling in potassium hydroxide to soften the tissues. The bases of the antennae are visible towards the top of the head (right), while towards the left are the numerous biting and seizing parts of the mouth apparatus, particularly prominent are the serrated mandibles (×15).

and allowing the stain to run off. The whole cell takes up the stain, but the nucleus appears much darker than the cytoplasm (fig. 33).

WHOLE MOUNTS OF INSECTS

Very small insects or crustaceans may be mounted whole if tissues are first softened (plate 3). This may be achieved by soaking specimen in 5% potassium hydroxide (KOH), which will eventually render the insect soft and fairly transparent, a process (clearing) which may take from several hours to several days, depending upon the robustness of the specimen. This clearing process can also be used for preparing parts of larger insects (mouthparts, legs, etc.) ready for mounting. The resulting specimen may be mounted unstained, or may be stained with borax carmine. For unstained mounting, the cleared insect, on removal from the potassium hydroxide solution, should be washed in water, dehydrated through an ascending series of alcohols, and mounted (fig. 34). If the cleared specimen is to be stained, immerse it in borax carmine for 10 minutes, then wash it in acid alcohol until it becomes bright red. Once again the time required for this final process is variable. Once the staining process is complete, the specimen may be dehydrated and mounted as described previously (fig. 35).

MOULDS AND OTHER FUNGI

Fungi are easily obtainable from decomposing organic matter. A moistened slice of bread, for example, kept in a warm, dark, humid place will soon become covered by moulds of a number of different species, and these are amenable to staining with lactophenol cotton blue. The method is especially suitable for parasitic fungi, since the fungal tissues will be preferentially stained (fig. 36). Lactophenol cotton blue can be a semi-

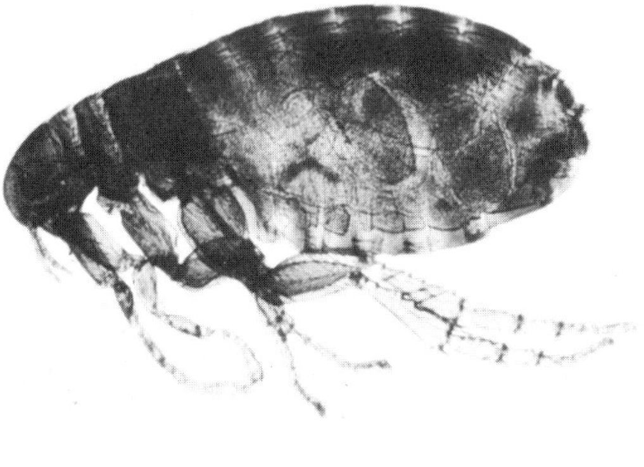

Fig. 35 Whole mount of a female human flea *Pulex irritans* (×20).

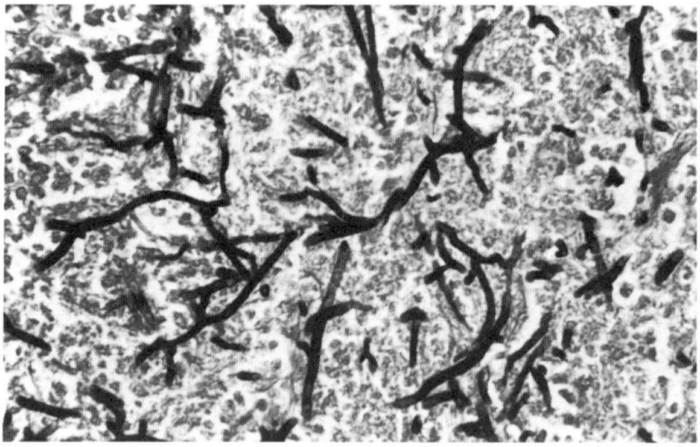

Fig. 36 The densely stained branching hyphae of the fungus *Aspergillus* in a human lung (×300).

permanent mounting medium, so all that is required is to place a small amount of the mould, or a section (see Chapter 4) of an infected plant directly into the stain on a slide, lower a coverslip and seal the preparation.

BACTERIA

Bacteria are visible under high magnification in the light microscope, but most of what is known about their structure has been obtained from electron microscope studies (see Chapter 9). Nevertheless, much of interest can be obtained by viewing bacteria in the light microscope, either as a living suspension, as described in Chapter 5, or else in conjunction with the staining techniques that have been developed for bacterial classification (fig. 37). The culturing and investigation of bacteria is an extensive subject in its own right, and the reader is referred to the bibliography for more information on this. Here, however, a widespread technique for staining and examining bacteria, Gram's method, is briefly described.

A scrape of plaque from the surface of the teeth provides a suitable source of bacteria for staining. The bacterial cells must be fixed, as a smear on the glass slide, prior to staining. This can be done by mixing the plaque sample with a small drop of sterile water on the slide. Stir the mixture so that the scraping becomes broken and evenly distributed over an area of about one square centimetre (fig. 38).

Allow the smear to dry in air. When it is dry, pass the slide through a bunsen, or spirit, burner flame three or four times to fix the cells, but do not allow the slide to get hot, as this will distort the cells. The smear is now ready for staining.

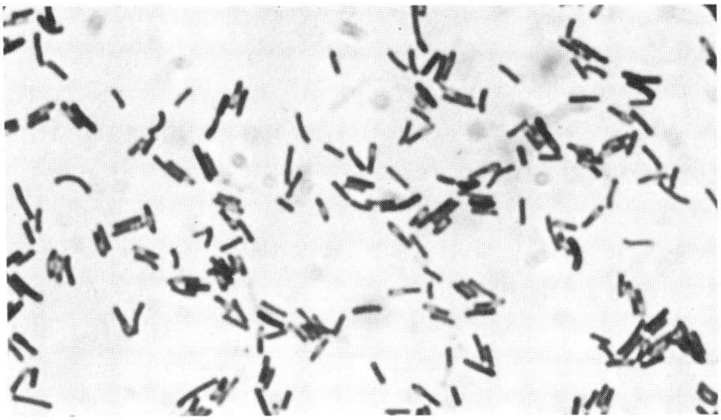

Fig. 37 Rod-shaped bacteria *Clostridium* sp. in a Gram-stained smear. Highly refractile spores can be seen inside some of the rods (× 1000).

Cover the smear with crystal violet stain, and leave for 5 minutes, after which the stain is tipped off, and the excess washed off with a small quantity of Gram's iodine solution. Some more iodine can be added to cover the smear, and left for 2 minutes. Drain the iodine from the slide, add acetone to the smear, and then immediately wash the acetone from the slide under gently running water. Staining is completed by applying basic fuchsin stain for 30 seconds, again rinsing in water, gently blotting, and then allowing to dry in air. Bacteria respond differently to this staining procedure, some becoming dark violet (referred to as Gram-positive), others pink (Gram-negative). This difference, which can be used for the preliminary classification of bacteria, is due to differences in the structure of the walls of these organisms.

BLOOD SMEAR

The major cellular constituents of blood, the red cells (*erythrocytes*) are clearly visible under the microscope without staining, but the use of a suitable stain helps to distinguish the white cells (*leucocytes*). Blood is generally mounted as a smear, and this is then suitable for a stain such as

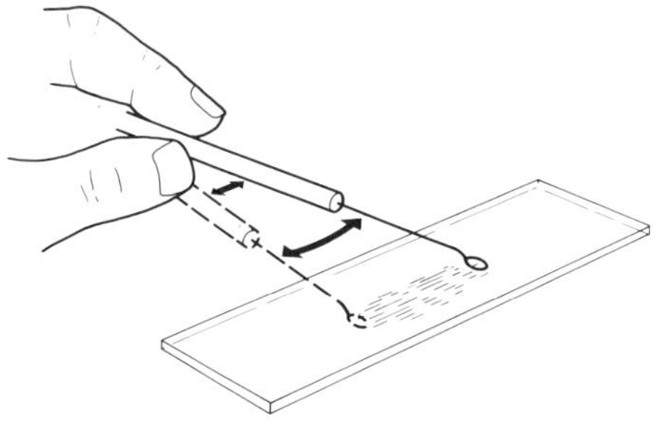

Fig. 38 Smearing a droplet of water containing bacteria, using a mounted loop of wire.

Leishman's, which is widely used to distinguish the different types of leucocytes.

A drop of blood is best obtained from the base of the thumbnail, preferably with a sterile disposable lancet, although a needle sterilised in a flame is adequate if it is discarded after use, since a re-used needle can lead to infection. Alcohol (70%) is suitable for sterilising the skin before and after piercing with the needle. Two slides are required for making the smear. The droplet of blood should be placed at the end of one of the slides. The other slide is then lowered until it touches the first slide at an acute angle, so that the blood is in contact with both slides. As the blood spreads along the line of contact the uppermost slide should be pushed rapidly along the lower slide, so that it draws the film of blood behind it, thus ensuring a thin uniform smear containing undamaged cells (fig. 39).

Once the smear has dried, apply sufficient Leishman's stain to cover the blood film, and leave for 2 minutes. Then add sufficient distilled water to cover the whole slide surface completely and leave for 15 minutes, during which time a greenish scum will form. Wash the slide in distilled water and allow it to dry in air. To make the slide permanent add a few drops of Canada balsam and a coverslip; or alternatively, for short-term examination, use 50% glycerine as described above.

Examination of the slide shows red cells to be predominant, but white

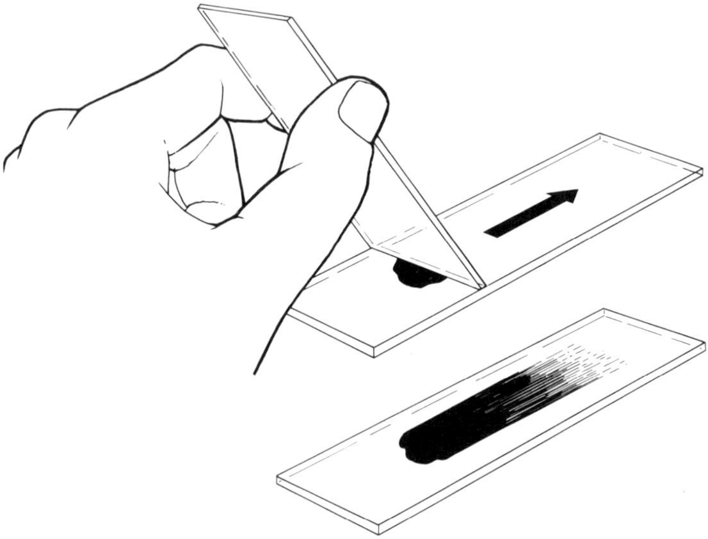

Fig. 39 Preparation of a blood smear. The rapidly moved slide pulls the blood behind it so that cells are not crushed.

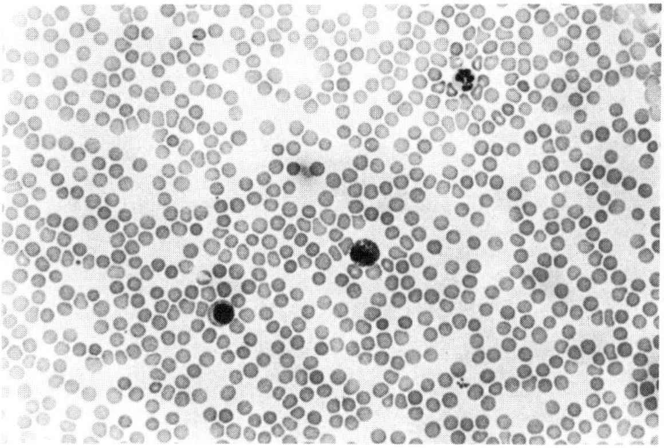

Fig. 40 Part of a stained human blood smear, showing vast numbers of enucleated red cells (erythrocytes), and three white cells (leucocytes) with stained nuclei (×300).

Fig. 41 A completed, labelled slide preparation.

cells should also be visible, with light blue cytoplasm and dark blue nuclei (fig. 40).

Labelling of Slides

Since it is not always possible prior to the examination stage to decide whether a slide is to be temporary or permanent, it is economical always to use a temporary labelling medium, provided that the one chosen will not be defaced by the particular staining or mounting procedure. Choice of labelling medium should therefore take into account the specific method of staining and mounting employed with each particular specimen. Temporary labelling media are felt-tipped pens and wax pencils (chinagraph). These can be removed by water or alcohol.

For slides which have been rendered permanent, the most satisfactory permanent labelling medium is an adhesive label (written in black ink to avoid fading). The label should be covered with clear varnish for waterproofing. It is a useful guide to the features highlighted in the specimen, if details of the particular stain and mountant used are also included on the label (fig. 41). The use of diamond pencils for labelling is not recommended, as they provide a poor visual display. The slides may then be stored indefinitely.

4
Plant and Animal Tissues under the Microscope

The examples discussed so far have involved small organisms mounted whole, or else tissues which are already in a form suitable for mounting (for instance blood). These need no elaborate pre-mounting treatment, other than the application of stains. Larger organisms, however, require a different sort of approach. This is usually achieved by *sectioning*, either of the whole organism or, more commonly, of specific parts of the organism. Sections are generally cut using a *microtome* (see below), although this is not always essential. Microtomes are designed to cut a ribbon of sections (serial sections), from which it is often possible to reconstruct the three dimensional structure of the specimen.

Sectioning Planes

The plane in which a section is cut is usually specified. The complexities of naming all possible planes are outside the scope of this book, so discussion will be restricted to the commonest planes; the *transverse section* (TS) and the *longitudinal section* (LS). The transverse section is cut at right angles to the long axis of the structure, whereas the longitudinal section is cut parallel to the longest axis (fig. 42). This may cause problems if several structures are present in different orientations. For example a TS of a plant root may well include an LS of a lateral root. In such cases it is customary to label according to the plane of the major structure, unless the section has specifically been cut for the minor structure. Obviously a longitudinal section may itself be cut in a number of different planes. For example, in a cylindrical object such as a plant stem, the plane of the LS may cut through a radius (a *radial LS*) or may avoid the centre entirely (a *tangential LS*, see fig. 42). In a flattened object, such as a leaf or the body of an animal, the longitudinal section may be vertical or horizontal.

Plant Tissues which do not require Sectioning

As stated above, techniques other than sectioning may yield cells or tissues in a form suitable for mounting. An example of this is the preparation of plant epidermal tissue. The outer covering of the non-woody parts

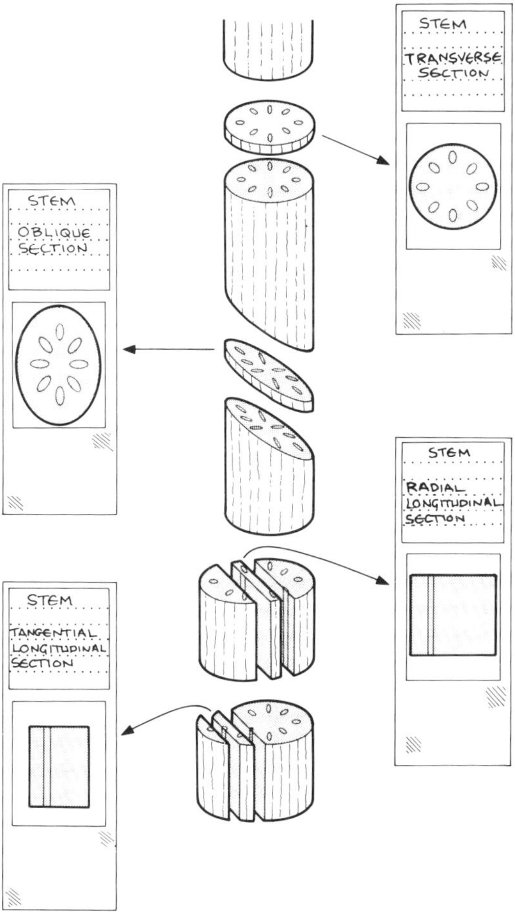

Fig. 42 Commonly used sectioning planes.

of a plant is a layer one cell thick. This epidermis may be obtained by mechanical or chemical means and examined in surface view (fig. 43). Patterns of epidermal cells are different in each species, so the technique may be used for identifying plant fragments. This has numerous uses, from the identification of plants in archaeological remains to the recognition by pharmacologists of plant residues as components of drugs.

Some epidermes are particularly easy to remove as whole sheets of cells. For example, if the fleshy leaves of an onion bulb are pulled apart, the thin transparent epidermis may be pulled from the inside of each leaf and mounted in water for examination. In general, however, a more elaborate treatment is required for removing epidermes from other plants. Often an epidermis can be exposed by scraping away the layers of tissue beneath it.

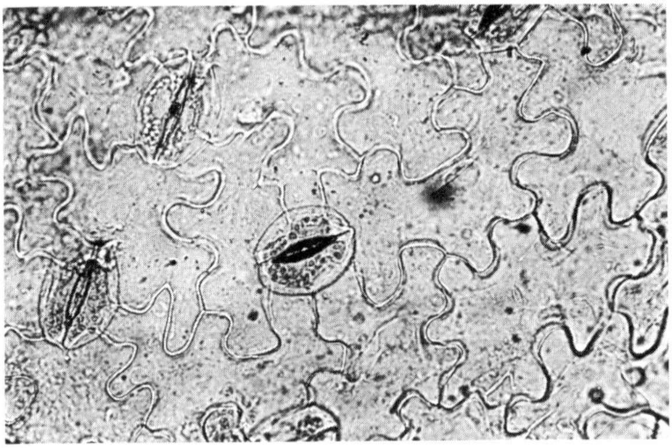

Fig. 43 Part of the epidermis of the celandine *Ranunculus ficaria*, showing cells with convoluted walls and irregularly spaced stomata with guard cells (×100).

This is best undertaken with a leaf, where the amount of tissue to be removed is fairly small. Place the leaf on a tile or some similar surface, and then scrape gently with a scalpel or sharp knife. Limit the scraping to an area of about 5 mm × 5 mm to avoid tearing the leaf. Scrape all tissues away to expose a small translucent patch of epidermis which can then be cut out and mounted in 50% glycerine, or dehydrated and mounted in Canada balsam. Since the pattern of cells is often different in the upper and lower epidermes, scraping can be undertaken from both sides of the leaf. This technique is ideal for the soft leaves of herbaceous plants, but is difficult with more robust shrub, tree and grass leaves, where the veins make scraping difficult.

In such cases *maceration* is an appropriate technique. Maceration involves boiling the tissue in a solution strong enough to separate the cells. For epidermal samples small pieces of leaf may be gently boiled in a dilute nitric acid solution, a method which has the disadvantage of producing noxious fumes. Gentle boiling breaks down the tissues between the upper and lower epidermes. Boiling time is variable: for a soft herbaceous leaf, 2 or 3 minutes is often sufficient, whereas a more fibrous leaf may take 15 minutes or more. The end point is reached when the leaf becomes clear and yellowish. It may then be removed from the solution with a fine brush, placed in water, and the two epidermes separated with a mounted needle. Again, either temporary or permanent mounts of the epidermes may be made.

Preparing Plant Sections by Hand

Firm plant structures, such as stems, can be conveniently sectioned by hand. A sharp cut-throat razor is best for this procedure; and though modern safety razors may also be used, they are more awkward to manipulate. For best results, hold the object to be sectioned between the finger and thumb of one hand. Hold the shaft of the razor in the other hand, between the thumb and the first two fingers; such that the sharp edge of the blade faces the operator, while the handle, gripped between the third and fourth finger, points away at right angles to the blade (fig. 44). Using this method of sectioning, the operator's thumb is at risk of being cut by the blade, and care should be taken to ensure that the thumb is always at least 10 mm below the surface to be sectioned. The razor should be drawn gently across the surface of the object towards the operator, to produce a series of transverse sections, the thinnest of which should be selected for staining and mounting. Often, taking incomplete sections rather than full ones can yield thinner material for mounting. Some softer plant specimens (for instance, leaves) may also be sectioned by hand if they are supported by placing them into a slit cut in a piece of elder pith wood or carrot, so that the specimen and the support material are sectioned together. In all cases, hand sectioning produces the best results when both the specimen and the razor are kept wet to reduce friction. Sections may be removed from the razor blade with a fine brush.

Longitudinal sections may be obtained in the following manner. A piece of the plant structure (eg stem) about 40 mm long is used. Two cuts are made, one on either side of the area to be sectioned, and each to a

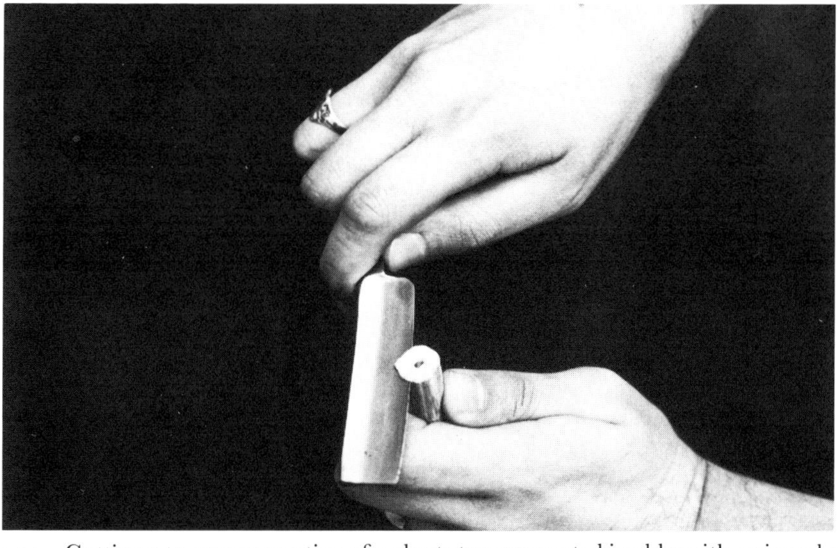

Fig. 44 Cutting a transverse section of a plant stem supported in elder pith, using a hand held razor. **Great care must be taken when using this method.**

depth of about two-thirds of the thickness of the structure. If the two end pieces are bent downwards at right angles, and the central part is supported over some suitable object, the sections can be cut by running the razor along the central part (fig. 45).

Vital Staining of Temporary Plant Mounts

Plant tissues and sections may be stained by any of the following methods, though none of these stains are suitable for permanent mounting. Methods for the staining of permanent mounts are given in a later section.

Most stains for plants are specific for constituents of the cell wall (fig. 46). The major component of the wall is cellulose, and this can be stained using Schultze's solution. With a brush, transfer the section to a few drops of Schultze's solution, stain for 5 minutes, and then mount in 50% glycerine. Cell walls containing mainly cellulose will appear blue, those containing deposits of other substances will appear yellow.

Many cell walls contain extensive deposits of lignin, and a number of lignin-specific stains are available. One of the most commonly used is aniline sulphate, in which the section is left for 5 minutes, transferred to 50% glycerine to wash out excess stain and then mounted in 100% glycerine. Aniline sulphate stains lignin yellow. Another widely used lignin stain is phloroglucin. Place the section in phloroglucin, leave for 5

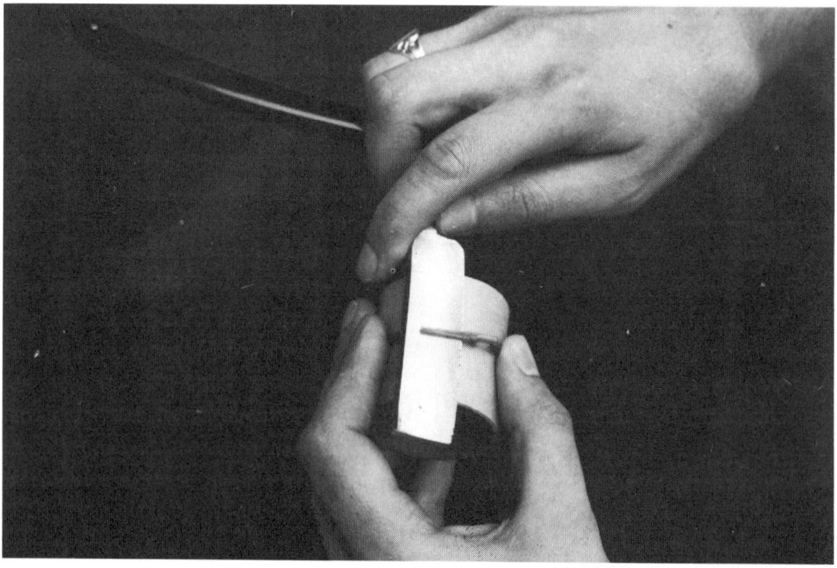

Fig. 45 Cutting a longitudinal section of a plant stem wrapped over a cylindrical object.

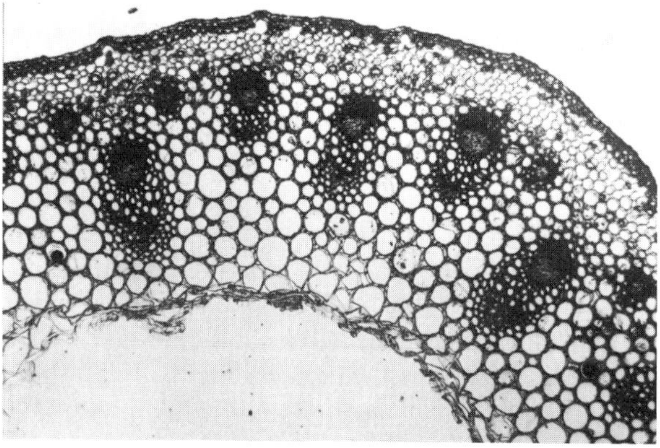

Fig. 46 Part of a transverse section of a deadnettle *Lamium* sp. stem which has been stained in aniline sulphate. The stain appears dark in cell walls in which thickening with lignin has occurred, as in the cells of the vascular bundles which are particularly prominent here (×50).

minutes, then wash section in a few drops of dilute hydrochloric acid. The stained section may be mounted in 50% glycerine, the cell walls containing lignin are stained violet.

Preparing Sections of Soft Plant and Animal Tissues

These require more elaborate methods than firm plant tissues, because the material needs supporting during sectioning. The preparation of sections from soft tissue involves three main stages: *fixation*, *embedding* and *sectioning*.

Fixation is the process whereby the cells of the tissue are preserved in as lifelike a state as possible. The fixative fluid kills and preserves the cells of the tissue so that no further change or deterioration can occur. Fixatives also maintain a chemical concentration similar to that of the tissue, so that shape distortion due to loss or gain of fluid is prevented. Material placed in the fixative should be small in size, preferably not more than 5 mm × 5 mm × 5 mm, to ensure adequate penetration of the fixative to all parts of the tissue.

Formol saline may be used for fixation of animal tissues, and formalin-aceto-alcohol may be used for plant tissues. Material should be left in the fixative for at least 6 hours (more for a large tissue), although tissues can be stored in it indefinitely, until required for further treatment. The volume of fixative used should greatly exceed the volume of the sample being fixed. Since the embedding medium is usually wax, which will not penetrate a specimen containing water, the fixed tissue must be dehydrated. The fixative itself (which is water based) must also be removed, and this is done by washing the specimen for several hours in running water. Dehydration can then be carried out in an ascending series of alcohols in the normal manner, except that, because of the size of the specimen, the time spent in each alcohol should be increased to 1 hour. To ensure thorough dehydration and consistent results, three immersions in absolute alcohol are required. However, the specimen should not be left in the absolute alcohol longer than recommended or it will harden.

Embedding is the process in which a suitable supporting medium is infiltrated into the tissue. The most widely used embedding medium is paraffin wax which, when molten, is infiltrated into the specimen and then solidified into a supporting block. The absolute alcohol however must be removed from the specimen by means of a *clearing agent*, which must be miscible (able to be mixed) with both the alcohol and the paraffin wax. Suitable clearing agents are xylol, clove oil, cedarwood oil and pink paraffin. The specimen should be immersed in the clearing agent twice, each time for 1 hour. The tissue can then be placed in molten paraffin wax for 1 hour. The wax will penetrate to all parts of the specimen. Particularly delicate specimens should first be immersed in a 50/50 mixture of the clearing agent and wax for 30 minutes. Transfer the specimen to a small container and cover with fresh molten wax. At this stage the orientation of the tissue in the wax should be noted, and altered if necessary, by means of a warm needle. Allow the surface of the wax to solidify and then plunge the container into cold water. The rapid cooling of the block of wax containing the tissue is essential if good quality sections are to be produced.

Prior to sectioning, excess wax should be trimmed off the block containing the embedded tissue. The block is then attached to a microtome, and a series of sections cut. The complete procedure for fixing and embedding is presented in table 3.

1	Depending upon size of specimen, leave tissue in fixative for 6–12 hours.
2	Wash specimen in running water for 2 hours.
3	Place specimen in 50% alcohol for 1 hour.
4	Place specimen in 70%, 90%, 95%, and 100% alcohol for 1 hour each time.
5	Place specimen twice in fresh 100% alcohol for 1 hour each time.
6	Place specimen in clearing agent for 1 hour.
7	Place specimen in fresh clearing agent for 1 hour.
8	Place specimen in molten wax for 1 hour and keep molten in a warm oven.
9	Place specimen in fresh molten wax and keep molten for 1 hour.
10	Place dish of wax with specimen into cold water to solidify wax.

Table 3. Schedule for fixation and embedding of tissues prior to sectioning.

The Microtome

The microtome is a device for producing thin sections, normally 5–10 μm, from a specimen embedded in a wax block. The block is attached to the microtome and advanced by known increments towards a steel cutting edge, a procedure that leads to the production of a ribbon of sections. Individual sections or groups of sections can be easily removed with a fine brush. A commonly used commercial microtome is shown in fig. 47.

The commercial microtomes used in histology laboratories are expensive, but cheaper hand microtomes can be bought. They can also be easily constructed. A large nut and bolt, with a fine thread, is all that is required. The wax block is placed within the cavity produced by unscrewing the bolt most of the way out of the nut (fig. 48). A razor or razor blade moved across the surface of the nut will produce sections, the thickness of which can be adjusted by advancing the bolt so that the wax block protrudes slightly above the surface of the nut.

Mounting and Dewaxing the Section

The action of the microtome often causes compression in the wax section which therefore requires extending. This is accomplished by placing the section in a small drop of water on a clean slide, which is heated gently. Remove excess water with blotting paper and place the slide overnight in an oven at 30 °C. This process should firmly attach the section to the slide. However, if the section fails to stick to the glass, an adhesive will be necessary. A suitable adhesive is Mayer's albumen. A small quantity of the adhesive is rubbed evenly over the slide, then the section placed on it in a small drop of water and dried as above.

Before staining can be attempted, the wax must be removed from

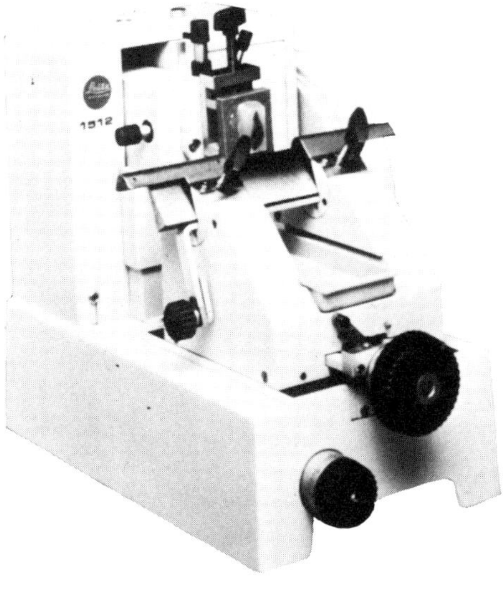

Fig. 47 A rotary microtome, the Leitz 1512. The embedded specimen can be seen attached to the block above the cutting edge.

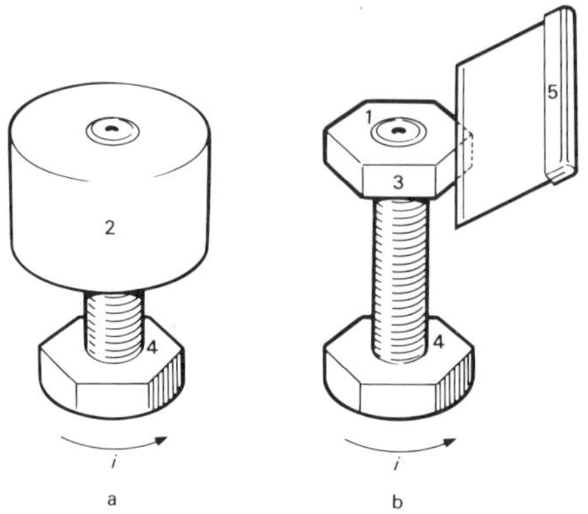

Fig. 48 (a) Example of a simple commercial hand microtome. (b) A home-made hand microtome. (1) Specimen in wax block. (2) Brass block. (3) Fine threaded nut. (4) Fine threaded bolt. (5) Single-edged razor blade. (*i*) Incremental advancement of wax block.

around the specimen, by placing the slide (for a few minutes) in the clearing agent which was used during the embedding of the tissue. The section can now be stained.

Staining of Animal and Plant Sections for Permanent Mounts

One of the stains most widely used for animal sections is the combination haematoxylin and eosin. This is an example of a *counter-staining technique* where the two stains are taken up by different parts of the section, and thus their colours contrast with each other. Haematoxylin is applied in water, and the section must, therefore, be rehydrated by passing the slide through a descending series of alcohols to distilled water, with about a minute spent in each. The section is stained for 5–15 minutes in haematoxylin and then washed in running water for 10 minutes, until the section is blue. Eosin is alcohol based, so the section must be dehydrated again to 90% alcohol before immersion for 2–5 minutes in this stain. Excess stain is removed by washing the slide in 90% alcohol, which results in the section being stained to the correct intensity. The time required for this process (*differentiation*) can only be adequately assessed by trial and error. The slide preparation can now be completed in the normal way by placing it in absolute alcohol, then xylene; finally add Canada balsam and a coverslip. The procedure for haematoxylin and eosin staining is displayed in table 4. With haematoxylin and eosin, nuclei are stained blue and cytoplasm pink. Fig. 49 and plate 1 show a section of animal tissue stained with this combination.

Plant material can also be stained with haematoxylin and eosin, although an alternative is safranin and light (or fast) green (table 5). Again, this is a counter-staining procedure, where differentiation is

1 Trim wax block containing specimen.
2 Cut section on microtome.
3 Mount section in a drop of water on a slide.
4 Heat the slide gently to stretch the section.
5 Remove water by gently blotting the slide at the side of the section.
6 Leave slide to dry overnight in a warm place.
7 If section fails to adhere stick it to slide with Mayer's albumen.
8 Remove wax from section by placing slide in clearing agent for 3 minutes.
9 Place slide in 100% alcohol for 1 minute.
10 Place slide in 70% alcohol for 1 minute.
11 Place slide in distilled water.
12 Place slide in haematoxylin for 5–15 minutes.
13 Wash slide for 10 minutes in running water until section is blue.
14 Place slide in 70% alcohol for 1 minute.
15 Place slide in 90% alcohol for 1 minute.
16 Place slide in eosin for 2–5 minutes.
17 Differentiate by washing section in 90% alcohol until stain intensity is satisfactory.
18 Place slide in 100% alcohol for 2 minutes.
19 Place slide in clearing agent for 2 minutes.
20 Drain slide, place a drop of Canada balsam on section, lower coverslip, and dry in oven.

Table 4. Schedule for staining sections with haematoxylin and eosin.

accomplished by adding excess stain, and then washing this out until a satisfactory intensity of colour is achieved. However, the dewaxed specimen only needs rehydrating down to 70% alcohol, and is then stained in safranin for 10–15 minutes. This is followed by washing in 70% alcohol, differentiation in acid alcohol, further washing in 70% alcohol, and dehydration in absolute alcohol. At this stage, the section may be counter stained in light green stain and differentiated in clove oil. The slide can then be finished in the usual way by immersing in xylene. Finally lower a coverslip on to a drop of Canada balsam. The combination of safranin and light green, stains the lignin of the cell walls red, and other parts of the cell green. Fig. 50 and plate 2 show stained plant sections.

1–10 As for table 4.
11 Take slide from 70% alcohol and stain in safranin for 10–15 minutes.
12 Wash slide in 70% alcohol until no more stain is removed.
13 Differentiate by washing slide in acid alcohol.
14 Place slide in 70% alcohol for 2 minutes.
15 Place slide in absolute alcohol for 2 minutes.
16 Place slide in fresh absolute alcohol for 2 minutes.
17 Stain with light green for 1–2 minutes.
18 Differentiate section in clove oil.
19 Place slide in clearing agent for 2 minutes.
20 Place slide in fresh clearing agent for 2 minutes.
21 Drain slide, place a drop of Canada balsam on section, lower coverslip, and dry in oven.

Table 5. Schedule for staining sections in safranin and light green.

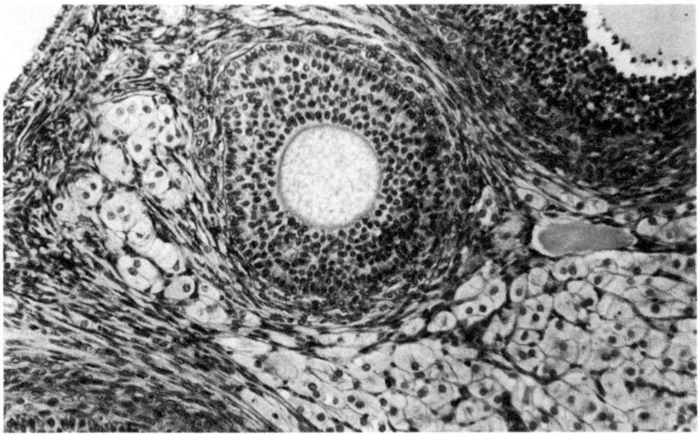

Fig. 49 Section of a developing follicle in a mammalian ovary, stained with haematoxylin and eosin (×250).

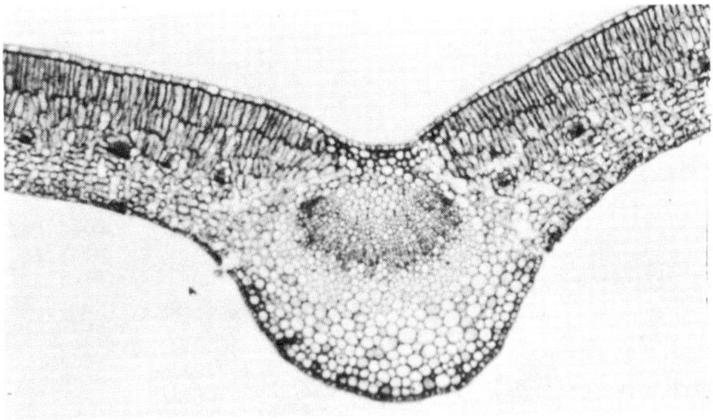

Fig. 50 Transverse section of the leaf of privet *Ligustrum vulgare* showing elongated palisade cells below the top epidermis, and numerous vascular bundles (veins) in the leaf (×50).

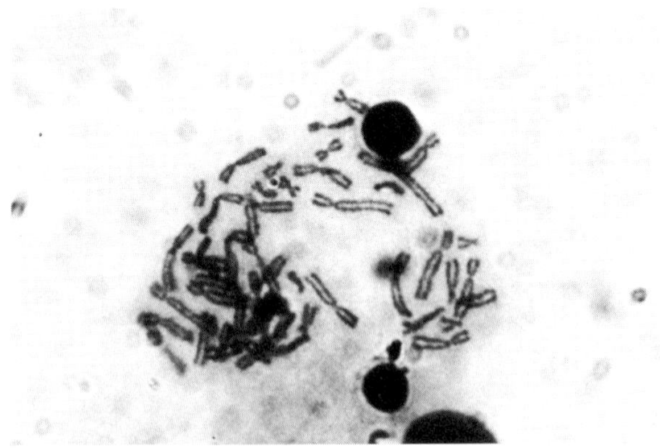

Fig. 51 A stained preparation of female human chromosomes obtained from cultured white blood cells. Each chromosome is longitudinally divided into two chromatids held together at the centromere, which appears as a constriction (×1500).

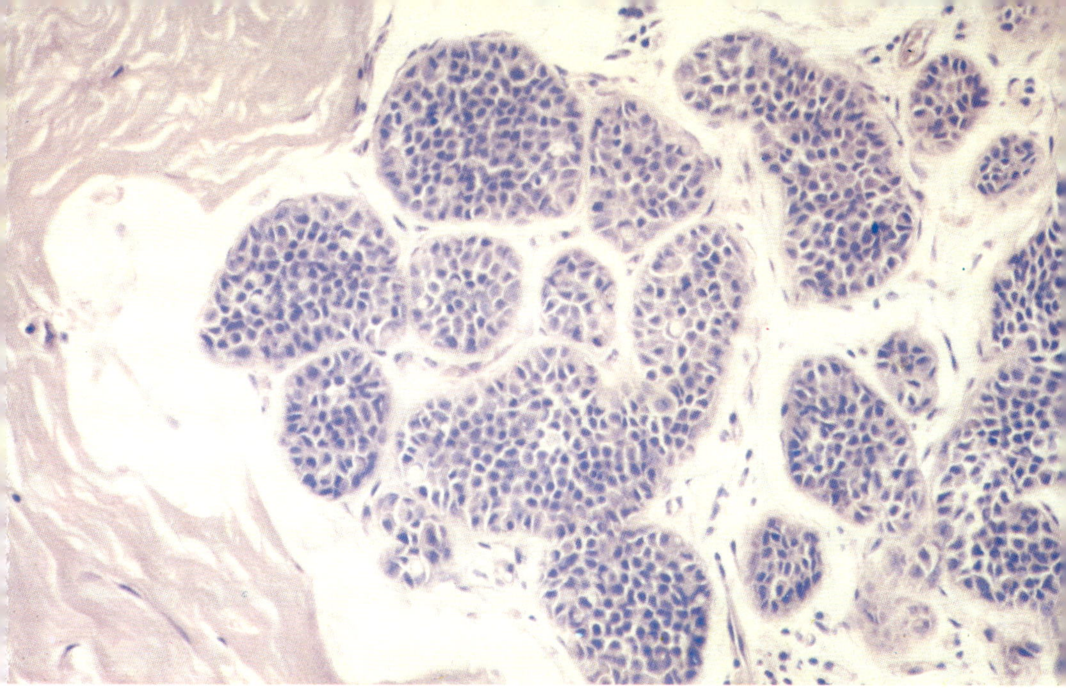

1 Haematoxylin and Eosin stained section of human breast, in which the ductules have become filled with malignant cells obliterating their lumina and expanding their diameter. This is a very early stage of breast carcinoma (*lobular carcinoma in situ*). In this stain combination Haematoxylin stains nuclei blue and Eosin stains cytoplasm and connective tissue pink (×200).

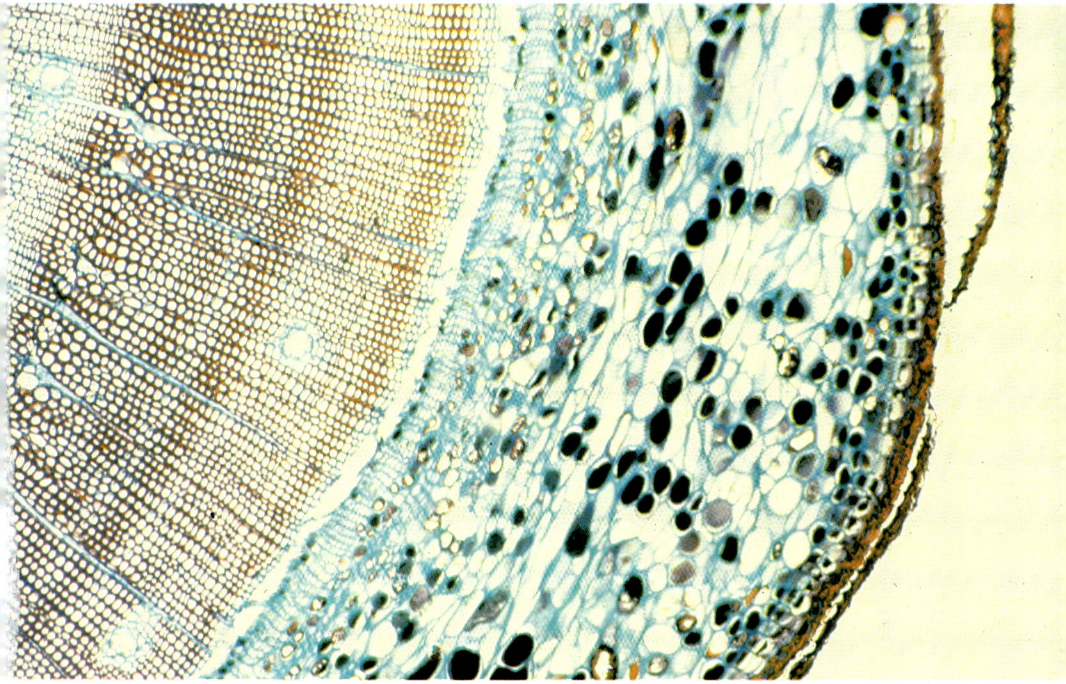

2 Transverse section of the stem of the pine *Pinus sylvestris* stained with Safranin and Light Green. Safranin stains lignin deposits (the basis of wood) red, as in the tissues on the left (×50).

3 Whole mount of the head of the male mosquito *Culex pipiens* showing the feathery antennae, the modified mouthparts for piercing skin (compare with the biting mouthparts of the cockroach in fig. 34) and the large compound eyes. (×25).

4 An igneous rock, eucrite, under crossed polars, showing interference colours. The striped areas have first order grey and whites typical of feldspars. The colourful areas are second order yellow, purple and violet, typical of augite when the section is the standard thickness of 30 micrometres. (×30).

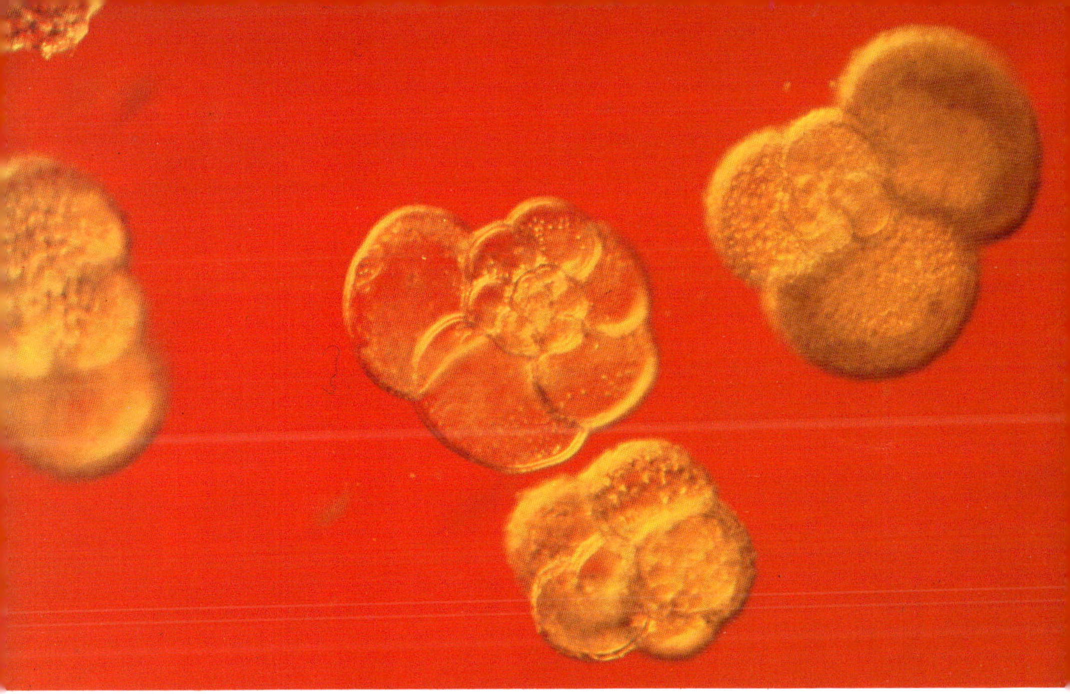

5 The foraminiferan *Globorotalia* in ooze from the floor of the Atlantic. The photograph was taken under Rheinberg illumination. (×200).

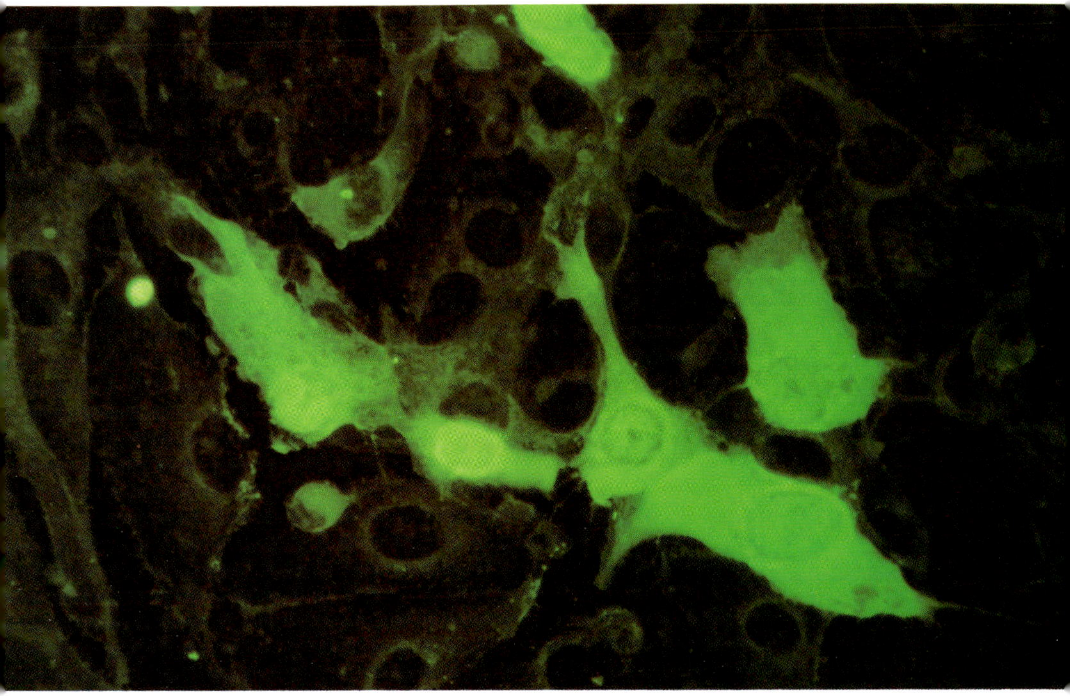

6 Fluorescent antibody stained influenza B virus in cultured cells, by the indirect method. (×400).

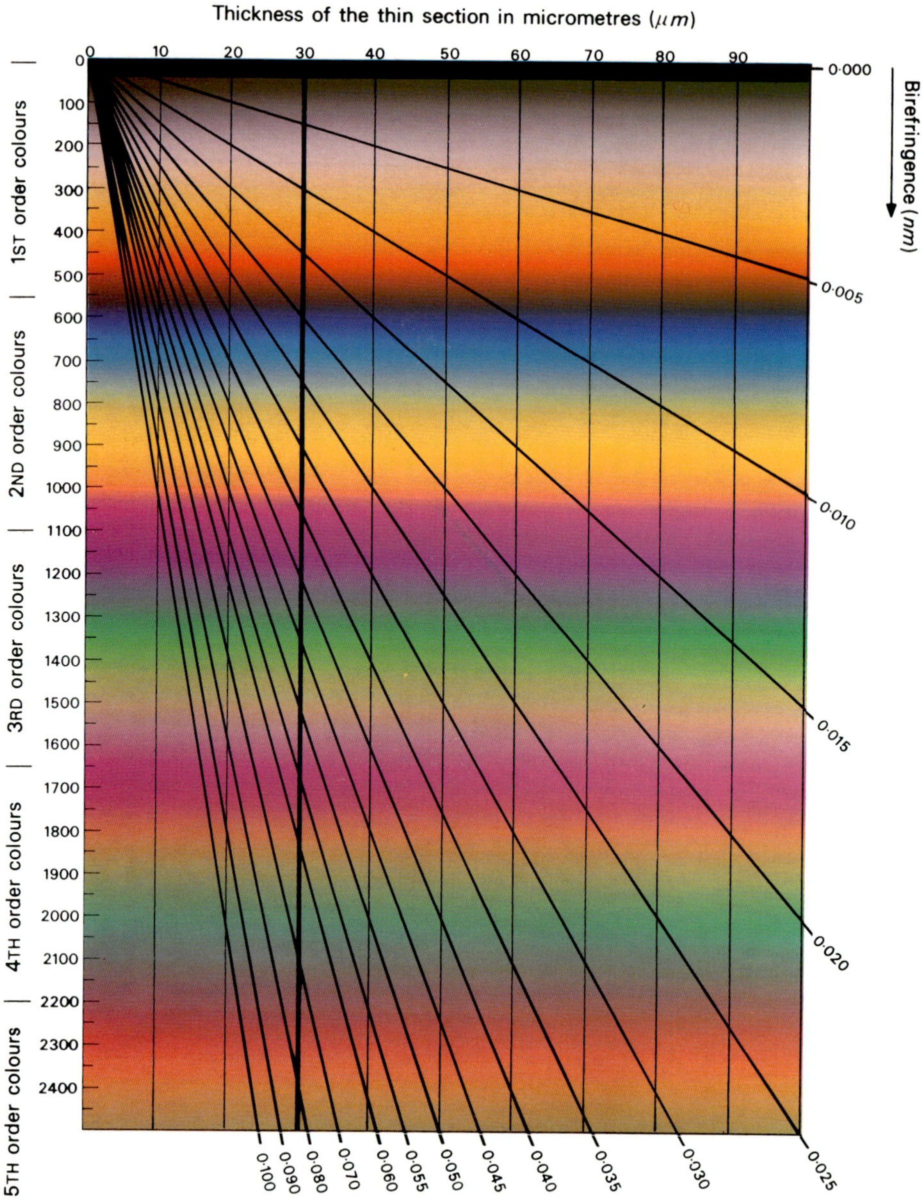

7 Michel-Levy Birefringence Chart illustrating the interference colours possible when crystals are viewed under crossed polarisers. The chart assists mineral identification — see both fig. 92 and chapter 6 for details.

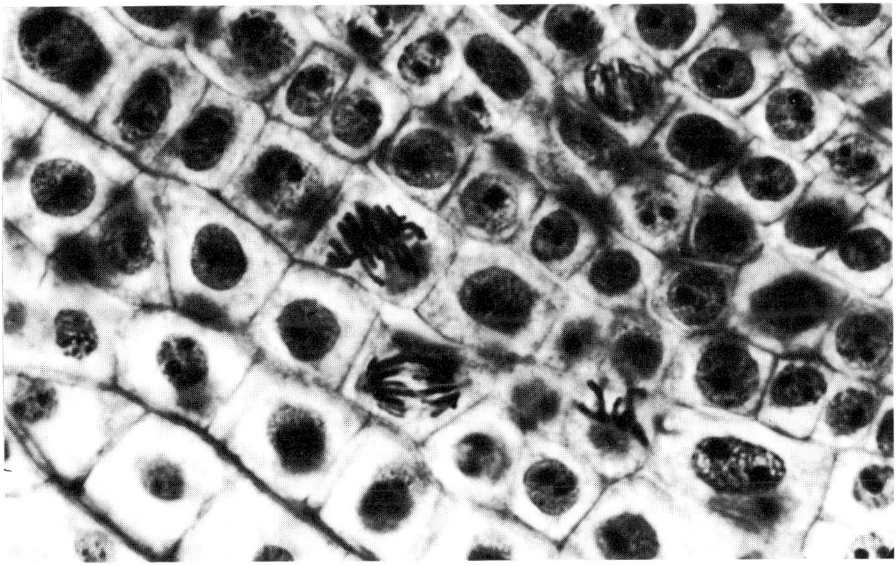

Fig. 52 Mitotically dividing cells in the root tip of the onion *Allium* sp. stained with acetic orcein. Various stages of mitosis can be seen. *Centre*: cell showing metaphase with chromosomes lined up in central plane of cell. *Lower centre*: cell showing anaphase, with chromosomes separating. *Upper right*: cell in prophase with chromosomes visible as long thin strands (×600).

Preparation and Staining of Dividing Cells

Dividing cells can be found in most tissues. They occur most abundantly at sites of active cell production. A growing root tip provides a good source of material for cells undergoing *mitotic division*. However, cells undergoing *meiosis*, which leads to ovum and sperm production, can only be found in reproductive tissues.

In both cases, the treatment of the cells involves the use of acetic orcein which stains the chromosomes, these being the most prominent feature of dividing cells (fig. 51). Cells representing all of the stages of mitosis or meiosis are detectable in these preparations (fig. 52). For a description of the various stages involved in cell division the reader should consult the sources listed in the bibliography.

Mitotically dividing cells can be examined in the tips of onion roots. Growth of these may be promoted by standing an onion bulb over water (eg on top of a bottle filled with water) in darkness for a few days, until roots 20–40 mm long grow into the water. Remove the roots and place them in acetic orcein and hydrochloric acid. Gently heat and leave for 15–20 minutes. Transfer a root from the stain to a drop of acetic acid on a

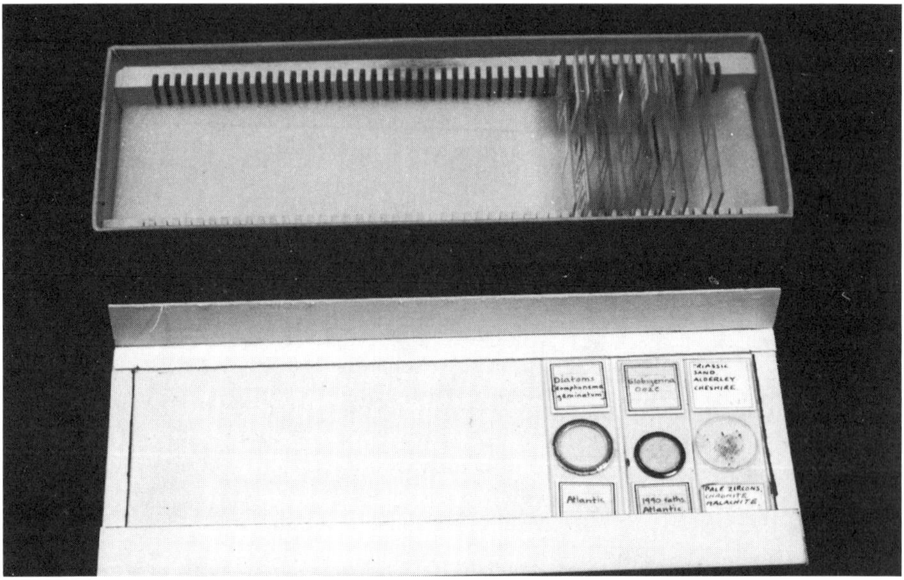

Fig. 53 A box and a tray for permanent storage of slide preparations.

slide and cut off and discard all but the terminal 1 mm of the root tip. Lower a coverslip over the root tip, and lightly tap the coverslip with a pencil point so that the cells separate and the whole root tip spreads out under the coverslip. Examination of the area just behind the tip apex will reveal the major zone of division, and chromosomes at various stages of mitosis will be found.

This stain and squash technique is also used for demonstrating cells dividing meiotically. The anthers of flowers are a suitable source of these cells, but are only seasonally available. They may, however, be stored in the fixative acetic alcohol until required for further treatment.

Storage of Slides

Permanent slides made by the methods described above may be stored in commercially available cabinets or in converted boxes. Two major types of cabinet are available: those with shallow drawers for lying the slides flat side by side, and those with deeper compartments and corrugated sides for storing slides vertically. Slide storage boxes can be constructed by lining the walls of a suitably sized box with corrugated cardboard. The slides can then be inserted vertically in the corrugations. Alternatively flat storage boxes can be constructed (fig. 53).

5
Aquatic Organisms under the Microscope

There is an enormous variety of animals and plants in aquatic environments. Many of these organisms, which belong to a number of different groups, are microscopic in size and provide a spectacular display when examined under the microscope. Aquatic organisms are easy to observe under the microscope since they require little preliminary treatment, and can be mounted directly in a drop of water on a slide. Both freshwater and marine organisms can be treated in this manner. Some additional techniques to facilitate the examination of these specimens will be discussed in this chapter.

Collecting Aquatic Organisms

Collecting samples of water from ponds, pools or slow streams is one of the simplest ways of finding aquatic microscopic organisms. The samples can be transported in clean screwtop jars, or glass or plastic tubes with caps. These can be carried in a polythene sandwich box, lined with expanded polystyrene suitably shaped for holding the containers steady. Many pond dwelling microscopic animals live on and around submerged aquatic plants or under stones, so searching these can be fruitful.

To collect samples from deeper submerged plants, a grapple is needed. This can be made from long nails bent to form a large hook, affixed to a beaten lead base and attached to a stout rope. The weeds, stones and water samples obtained should be left to stand in a white tray long enough for small organisms to emerge from hiding. A ×10 hand lens is useful in searching for small animals and plants. As most are sensitive to heat; samples collected from ponds should be kept cool. Specimens to be transported are unlikely to survive unless kept in a cool box or thermos flask. If destined for permanent mounting they can be preserved in 70% alcohol.

A variety of small organisms live in the waterlogged mosses found in swampy areas, especially acidic *Sphagnum* bogs. Some wet *Sphagnum* can be collected with a little water from seepage, and the drops squeezed from this examined under the microscope.

Samples of sea water or pond water are likely to contain microscopic

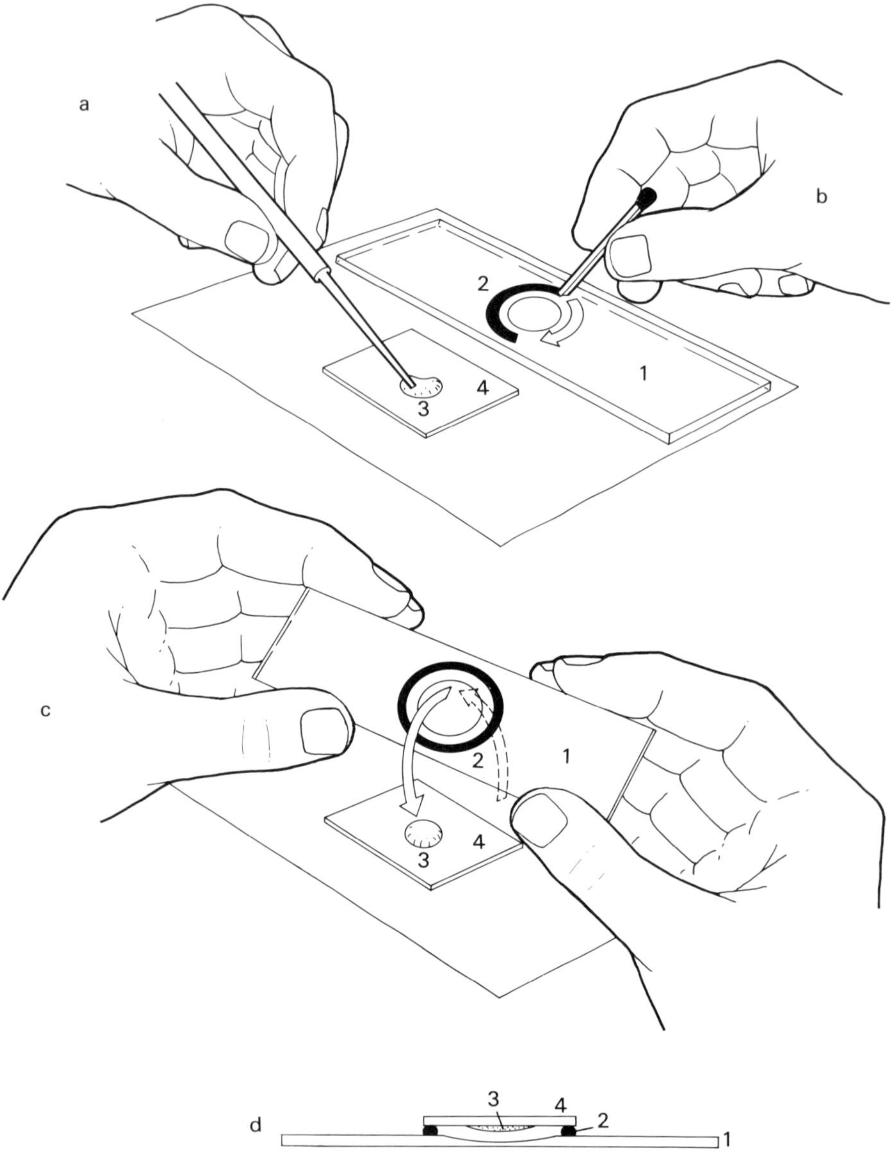

Fig. 54 Preparation of a hanging drop mount. A cavity slide (1) and a coverslip (4) are placed on a sheet of tissue paper placed on a table. Then (a) transfer a drop of the infusion or culture water (3) on to the coverslip using a mounted needle. (b) Apply a ring of vaseline (2) around the cavity of the slide using a clean matchstick or glass rod. (c) Invert the cavity slide and lower it on to the coverslip, then press down gently to make an airtight seal. (d) Reinvert the cavity slide and examine under the microscope.

organisms in suspension and free swimming (collectively called *plankton*). These are better collected with a net, which concentrates them into a smaller volume. The size of the mesh is important, since a fairly coarse mesh will trap the animals (zooplankton), while most of the small plants (phytoplankton) will pass through. Zooplankton and phytoplankton can therefore be collected separately by using two nets of different mesh size, and meshes of approximate aperture 0·3 and 0·1 mm respectively are suitable for this. Plankton nets may be purchased from suppliers or may be homemade from nylon or muslin mesh (see Appendix II for suppliers). The net may be mounted on a pole or towed slowly behind a boat.

OTHER METHODS OF OBTAINING AQUATIC ORGANISMS

Another method of obtaining microscopic aquatic organisms is by setting up a hay infusion. To do this, chop up a little grass and put the pieces in a screwtop jar, three-quarters filled with water. Lightly screw on the lid and store in the dark. The resulting broth is capable of supporting a growth of bacteria and protozoa present on the grass and in water. Other shredded plants can be used instead of grass. Additional nutrients can be released if the pieces are boiled in the water, and subsequently removed to leave an enriched medium, but in this case boiling may have sterilised the medium so the organisms must be deliberately introduced. This is easily accomplished by adding a drop of pond water, a speck of soil or a piece of unboiled plant, which will harbour bacteria, protozoa and algae. Unfortunately, some of the organisms may be injurious to health, so precautions should be taken. The culture should be stored away from food, preferably outside; hands and the glassware used should be thoroughly washed after handling the infusion, and all spillages should be mopped up with disinfectant; contaminated fingers should not come into contact with the eyes, nose or mouth while an infusion is being studied.

Live cultures of many species of protozoa and algae can be obtained from suppliers (see Appendix II). The cultures can usually be kept for some time if maintained in a suitable medium. Examples of commonly kept forms are species of *Amoeba* and *Paramecium*, and media for culturing these are given in Appendix I.

Special Mounting Methods for Aquatic Organisms

To examine living aquatic organisms under the microscope, a simple water mount may suffice as described in Chapter 3. However, problems of evaporation and overheating are then likely. Evaporation can be avoided by sealing the coverslip with vaseline, although this ultimately results in

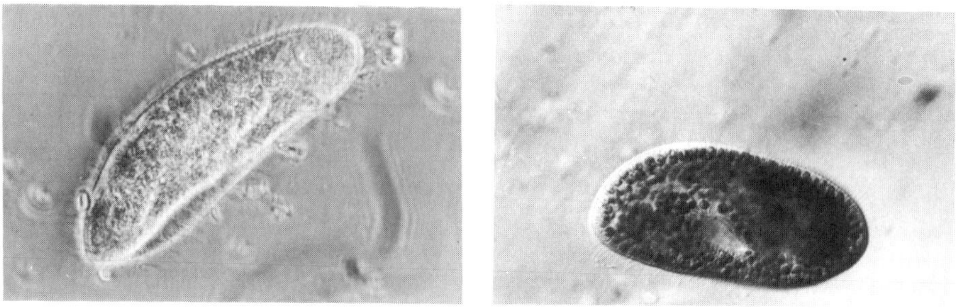

Fig. 55 (a) *Paramecium caudatum* (×200). (b) *Paramecium bursaria* showing large inclusions which are symbiotic green algae living within the animal (×200).

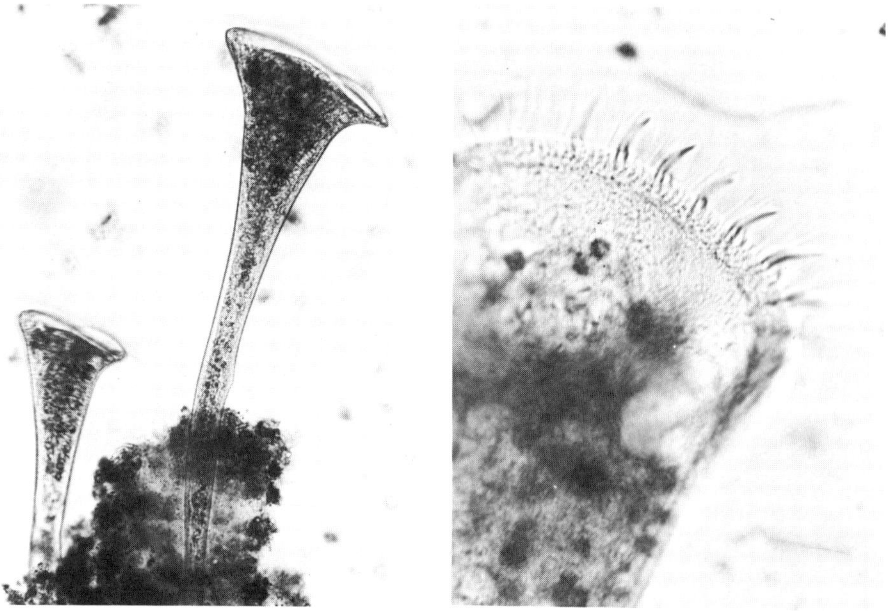

Fig. 56 (a) The ciliate *Stentor*, temporarily attached and feeding (×75). (b) Cilia creating a feeding current in *Stentor* by rhythmic (metachronal) beating (×150).

oxygen depletion in the water under the sealed coverslip.

An alternative is to prepare a *hanging drop mount*, in which oxygen depletion is less of a problem, and which provides a better opportunity for extended observation of certain organisms. A cavity slide or a slide with a raised ring cemented on to it is necessary for this technique. The steps involved in making a hanging drop mount are illustrated in fig. 54.

Many microscopic organisms move too rapidly to be kept easily in the field of view. A remedy for this problem is to add a small piece of cotton wool or a few strands of filamentous algae to the water, to trap the moving organism in a smaller space. Another approach is to add a 2% solution of methyl cellulose to the water, as this slows down the swimming organisms because of its viscosity.

Finally, actually seeing or finding some living microscopic animals, if almost transparent in transmitted light, may be a problem. Vital staining helps and is discussed below. Other remedies are described in Chapter 8.

The remainder of this chapter is concerned with examples of organisms which may be encountered in the samples collected.

BACTERIA

A rich growth of bacteria is often the most noticeable feature of a hay infusion. The bacteria form a scum over the surface, and also give the water a cloudy appearance. A drop may be examined in the usual manner under high magnification. In addition a bacterial smear can be prepared and stained by Gram's method, as described in Chapter 3.

PROTOZOA

Protozoa are abundant in pond water and hay infusions, or may be purchased as pure cultures. It may be useful in some cases to stain the living organisms; and suitable vital stains for this are Janus green B, which stains mitochondria, and neutral red, which stains the contents

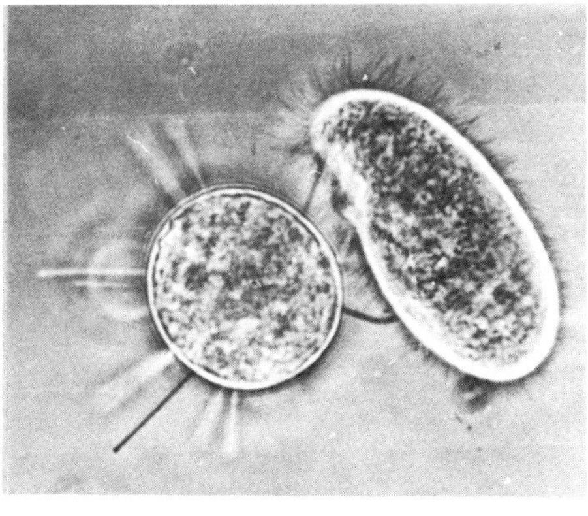

Fig. 57 The suctorian ciliate *Discophrya* feeding on an immobilised *Colpidium*. Note the tentacles used for feeding (×400).

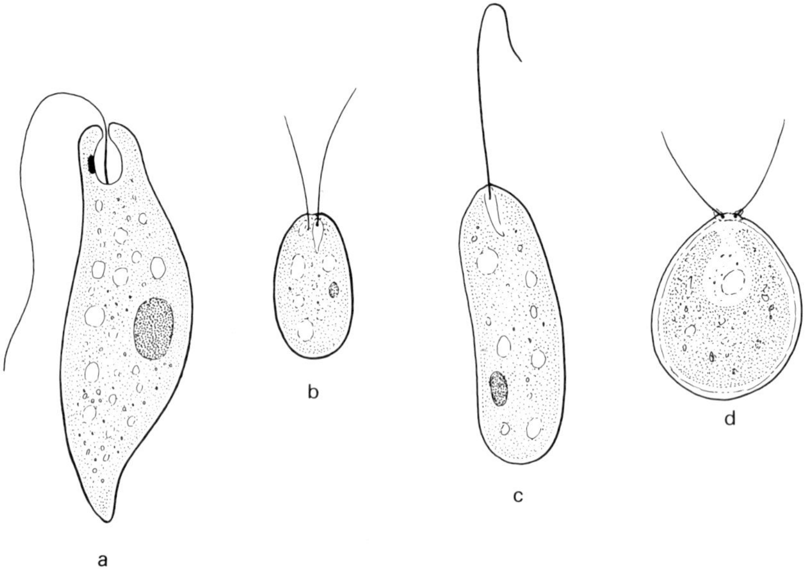

Fig. 58 Four common freshwater flagellates. (a) *Euglena* (×1000). (b) *Chilomonas* (×650). (c) *Peranema* (×600). (d) *Chlamydomonas* (×2500).

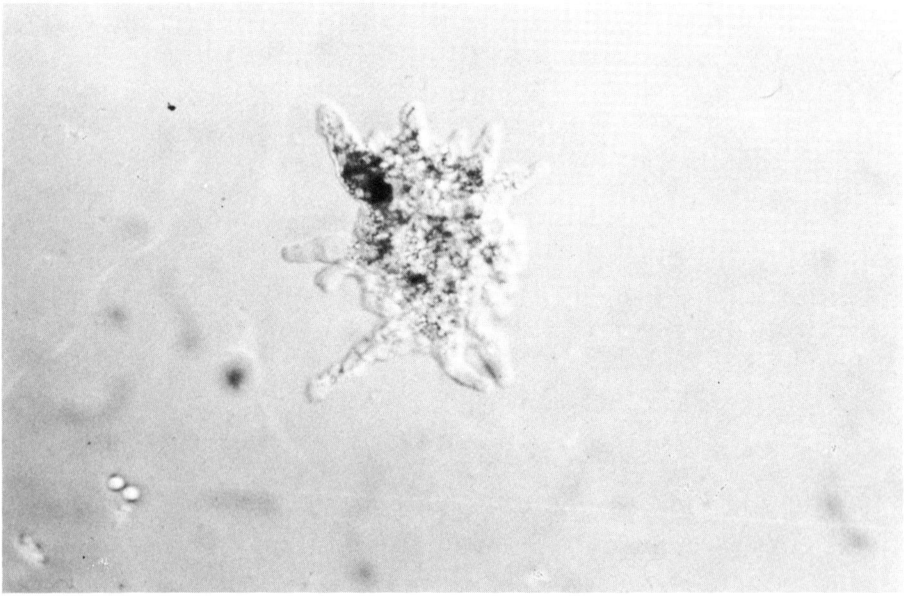

Fig. 59 The irregularly shaped *Amoeba proteus* (×90).

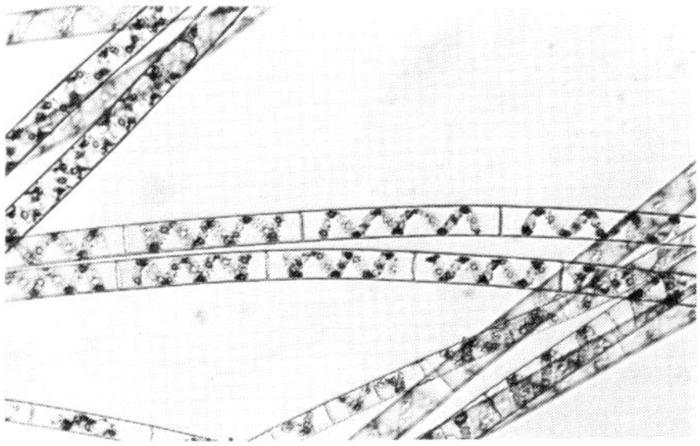

Fig. 60 The filamentous alga *Spirogyra* consisting of a string of cells, each with a coiled chloroplast (×75).

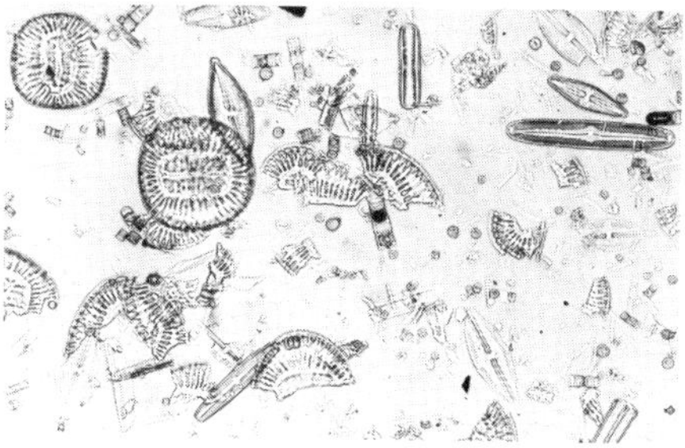

Fig. 61 Various forms of frustules in diatomaceous earth (×85).

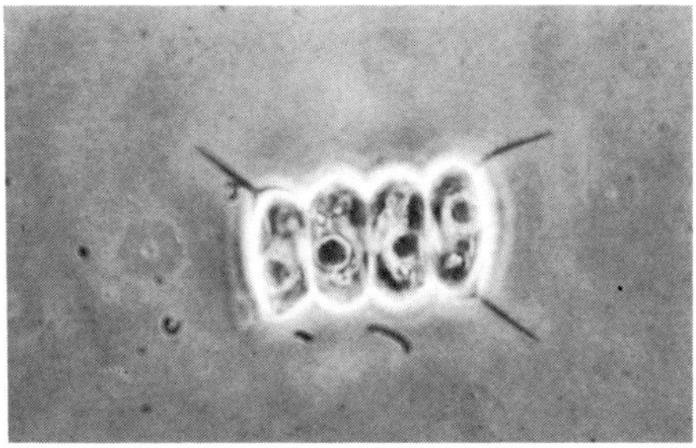

Fig. 62 *Scenedesmus*, a freshwater planktonic alga (×800).

of vacuoles. The stains should be very dilute (at least 1:10,000, see Appendix I) when added to the droplet containing the animals. Neutral red is a pH indicator, so stained food vacuoles show a colour change (red to yellow) as digestion proceeds. The main protozoans likely to be found are ciliates, flagellates and amoebae.

a) Diverse ciliates are found in hay infusions and often occur in vast numbers after a week or so. In particular, species of *Coleps*, *Tetrahymena*, *Colpidium* and the much larger *Paramecium* are likely to be present (fig. 55). *Paramecium* and *Colpidium* are good subjects for staining the food in food vacuoles with neutral red. Congo red stain may be used to colour food particles such as yeast cells, which can be fed to the ciliates. Purchase some live yeast, make a suspension of the yeast cells in water, and add to the stain. Leave for ten minutes until the yeast cells are red. Mix a drop of the stained yeast suspension with a drop of the *Paramecium* culture on a slide, and watch the ciliates feeding on and digesting the yeast. The pH change in the food vacuoles will turn the stain from red to a blue colour as digestion proceeds.

Attached ciliates such as *Vorticella* and *Stentor* (fig. 56), may also be found on submerged plants (collected by grapple as described above). The stalked *Vorticella* may also occur in colonies on the underside of the bacterial scum in an old hay infusion.

Small cultures of all these ciliates may also be purchased. The suctorian ciliate *Discophrya* (often erroneously referred to as *Podophrya*) can be obtained in this way and can be fed upon smaller ciliates such as *Colpidium* (fig. 57) and *Tetrahymena*.

b) Flagellates are usually found plentifully in pond samples and infusions. They have one or two whip-like flagella and often move rapidly. A possible stain for examination of flagellates is Lugol's iodine added to the droplet containing the flagellates on the slide. Lugol's iodine stains nuclei, any starch inclusions and often the flagella. Many flagellates contain chloroplasts and their resulting green colour renders them visible without staining.

Of the commoner green flagellates, *Euglena* and *Chlamydomonas* may be found in abundance in water obtained from pools in farmyards (fig. 58). A laboratory culture of these will thrive if a few drops of liquid plant fertiliser are added and the culture kept in daylight. A common colourless flagellate is *Chilomonas* (fig. 58), which is a suitable candidate for iodine staining, and for some of the special techniques described in Chapter 8.

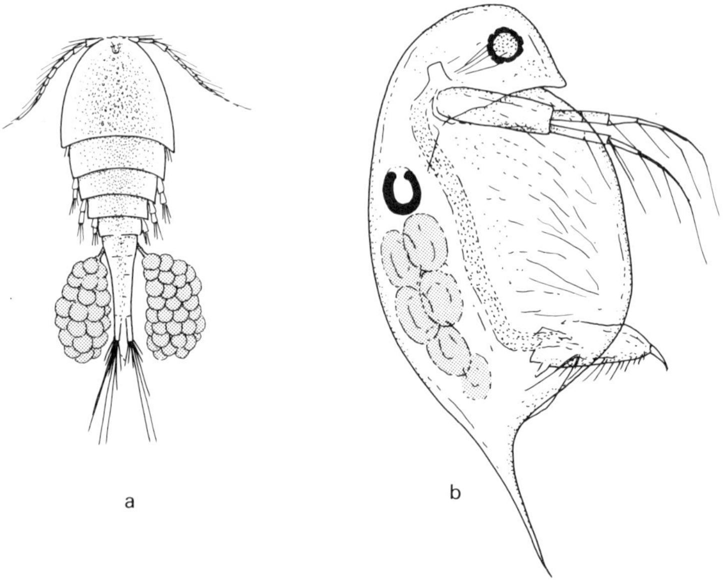

Fig. 63 Two freshwater crustaceans. (a) *Cyclops* with egg sacs (×20). (b) *Daphnia* (×30).

c) Several types of amoeba may occur in infusions and water samples. *Amoeba* itself (fig. 59), for example, is often found immediately below the bacterial film on a hay infusion, and a water sample pipetted from here is worth studying. In general amoebae are slow-moving and can be observed with ease under the microscope, although they are better seen by stopping down the iris diaphragm as described in Chapter 8, which will facilitate finding them on the slide. Shelled amoebae may be found in water samples from *Sphagnum* bogs, notably *Arcella* with its secreted shell and *Difflugia* with its shell of cemented sand and silt grains.

ALGAE

A number of different types of microscopic plants, all described by the general term algae, occur in water samples collected from ponds. They are also found in infusions kept in the light. One of the best known, is the filamentous *Spirogyra* (fig. 60) which forms slimy green masses in many ponds in summer, and which can be kept easily in laboratory culture if a few drops of concentrated house-plant food or a small quantity of soil is added to the water. Filamentous algae are also useful for oxygenating animal cultures and providing sites of attachment for sedentary animals, but must be prevented from overwhelming the culture.

Water samples, especially those from *Sphagnum* bogs, may contain large numbers of diatoms and green desmids. Both can be examined alive using the methods described above. However, for diatoms, permanent mounts reveal details of the *frustule* (siliceous exoskeleton) they secrete (fig. 61). Such mounts are made by placing a drop of water containing diatoms on a coverslip and laying the coverslip droplet uppermost on a strip of copper held by wooden tongs. If the strip is held in a bunsen flame, all organic matter will be burnt away, leaving just the pure white frustules attached to the coverslip. The heating should be gentle to avoid melting the

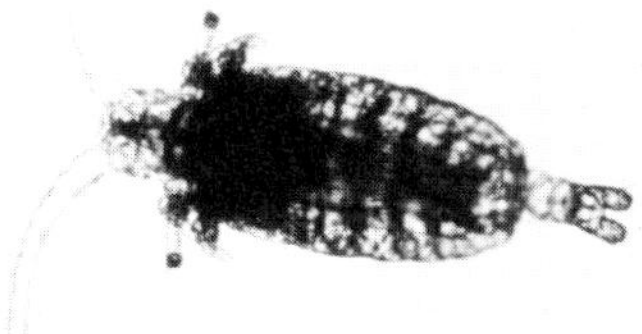

Fig. 64 A marine planktonic crustacean, *Calanus* (×15).

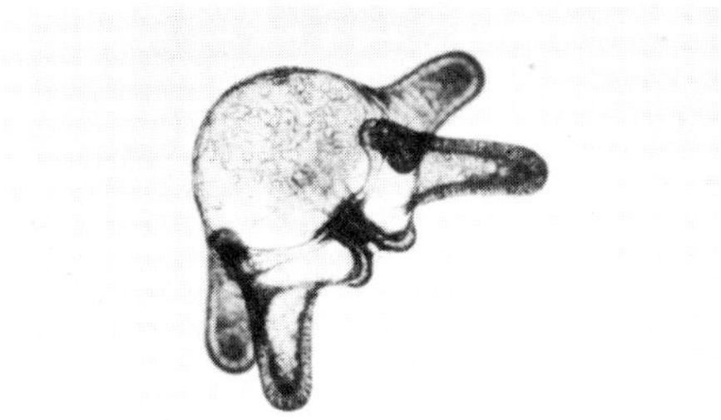

Fig. 65 The Müller's larva of a polyclad flatworm of the marine plankton (×15).

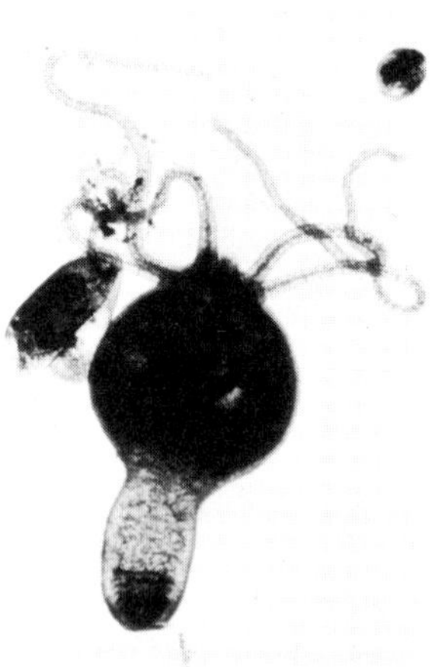

Fig. 66 A hydra, *Hydra oligactis*. The photograph includes a freshwater ostracod (*left*) and a chydorid (×6).

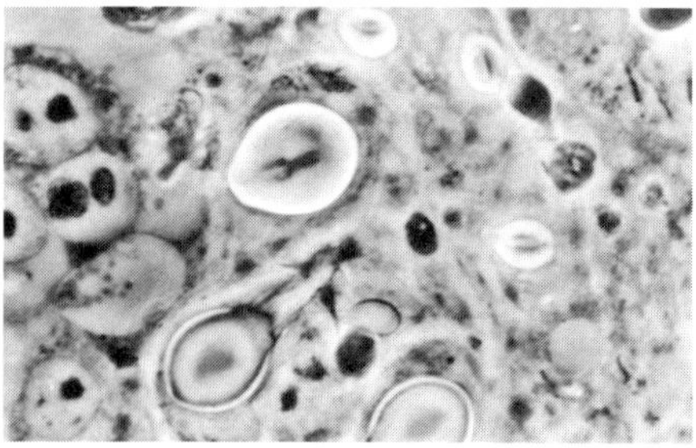

Fig. 67 Discharged and undischarged stinging capsules (nematocysts) of *Hydra*. Barbs are visible on the discharged nematocyst (×800).

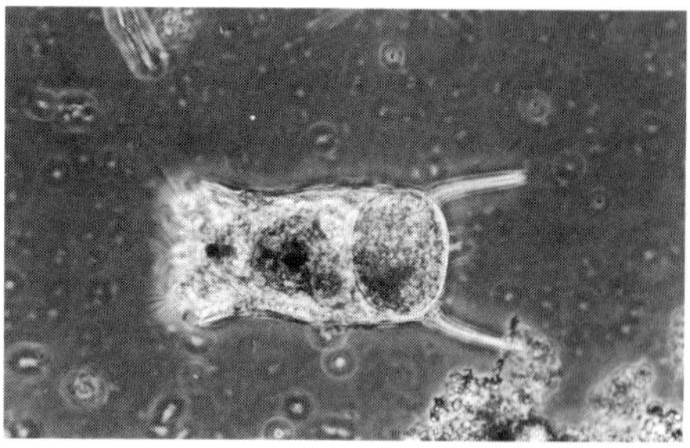

Fig. 68 *Brachionus*, a free swimming rotifer, showing ciliated corona (×70).

coverslip. The frustules are mounted in Canada balsam in the normal manner, although their visibility is better if they are mounted in a synthetic resin with a very high refractive index such as Naphrax. Invert the coverslip bearing the frustules and lower it gently on to a drop of Naphrax on a slide. Since Naphrax is in toluene solvent, the slide must be gently heated to dry the solvent, and then allowed to cool and set.

PLANKTON

Freshwater phytoplankton consists almost entirely of algae (fig. 62). Standing water may be temporarily coloured green by a sudden population explosion of the phytoplankton (a bloom). Among them may be found the colonial flagellated *Volvox*, which is large enough to be examined under low power.

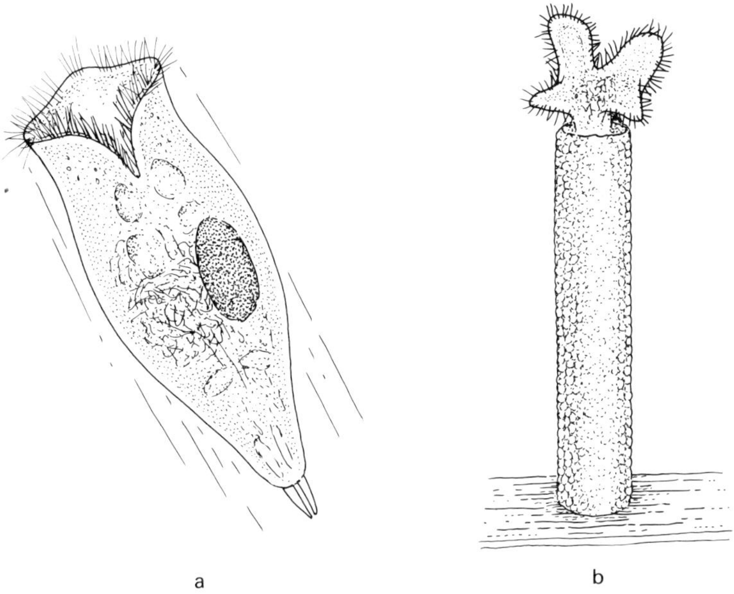

Fig. 69 Two rotifers. (a) The free swimming *Epiphanes* (×120). (b) The tube-dwelling *Floscularia* (×50).

Most prominent among the zooplankton of freshwater habitats are small crustaceans, which can be mounted and examined in water in the usual way, although a cavity slide or raised coverslip may be necessary. *Cyclops* and *Daphnia* are the best known of these crustaceans (fig. 63).

A much greater variety of crustaceans and other animals are found in marine zooplankton, which contains the larval stages of various invertebrates as well as adult crustaceans (figs 64 and 65). Unfortunately marine plankton deteriorates very rapidly (several hours) unless kept cool and well aerated. Transportation in a thermos or cool box is essential if the sample is to be carried some distance before examination, as this may extend the sample's viability by a few hours.

HYDRAS

These are best sought on the undersides of submerged stones or upon underwater vegetation. They may be particularly abundant in the outlets of lakes, where water currents remove the chemicals which each hydra normally secretes to prevent others attaching nearby. Since hydras grow larger at low temperatures, late spring and early autumn are the best times to search for them.

Cultures of hydras can be maintained more or less indefinitely in half-filled jars, provided they are kept cool (see Appendix I). The live animals may be observed by placing them on a slide with plenty of water, and then lowering a coverslip (fig. 66). Identification is difficult and requires examination of the stinging capsules (nematocysts, fig. 67) and details of the hydra's life history.

OTHER ORGANISMS

Microscopic animals of many other types are frequently encountered in pond samples and in weeds. Rotifers, for example, with their characteristic ciliated wheel organs (the corona), include free-swimming forms such as *Brachionus* and *Epiphanes*, as well as attached tube dwellers like *Floscularia* which may be found on submerged plants, along with creeping forms such as *Philodina* (figs 68 and 69). Also creeping are numerous other sorts of invertebrates, including crustaceans (ostracods and chydorids, fig. 70), tardigrades, gastrotrichs (also found in gutters) and flatworms. Encrusting organisms may also be encountered, including bryozoans, sponges, and the hydroid *Cordylophora* (fig. 71). Sponges, which are most abundant in dark, well aerated water, can be crushed into single cells and siliceous spicules in a droplet of water, and examined in a water mount.

Further details on the organisms mentioned in this chapter, together with keys for identifying them, are available in the publications listed in the bibliography.

FOSSIL AQUATIC ORGANISMS

The exoskeletons of many microscopic organisms accumulate into thick

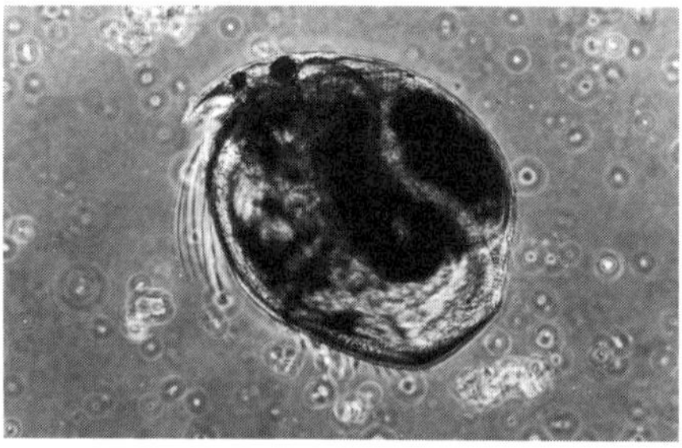

Fig. 70 A freshwater chydorid (×80).

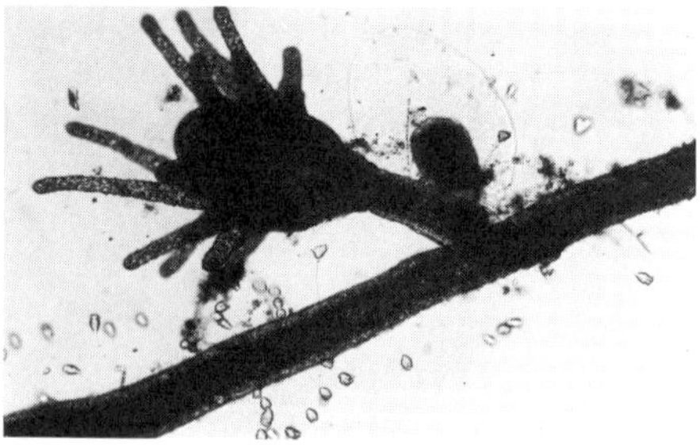

Fig. 71 A single hydroid polyp of the colonial *Cordylophora*. Numerous stalked *Vorticella* can be seen attached to the main stalk of the *Cordylophora* colony (×20).

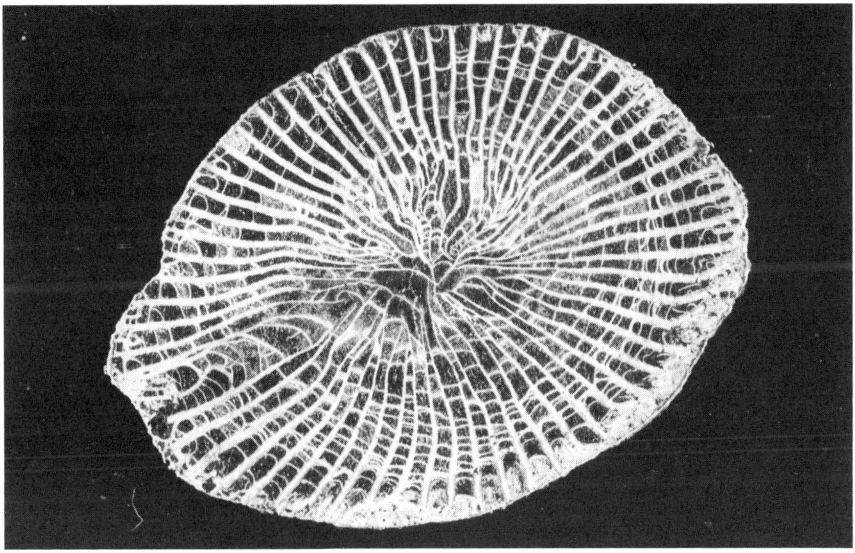

Fig. 72 An acetate peel from an etched fossil coral *Dibunophyllum* of the Lower Carboniferous (Mississippian) (×4).

sedimentary deposits, samples of which can be prepared for examination under the microscope. A good example is *diatomaceous earth* or *diatomite* (also known commercially as Kieselguhr), which consists of the frustules of diatoms, which are best mounted in Naphrax as described previously.

Some marine protozoa possess exoskeletons; the remains of these form radiolarian and foraminiferal oozes (plate 5). Some of these, such as Barbados earth, are obtainable from suppliers. Limestone may contain the remains of calcareous algae and foraminifera, rendered visible by rock sectioning, as described in Chapter 6.

Fossil corals are best examined in section by means of an acetate peel (fig. 72). To do this, fossil corals in limestone should be ground flat with a carborundum slip on a lapping plate as described in Chapter 6. Finish the surface with a wet cloth covered with mild abrasive paste (for example, domestic scouring powder) followed by application of cold, very dilute hydrochloric acid for 10 to 20 seconds. Once the surface is perfectly dry, the peel can be prepared as described in Chapter 3.

6

Minerals, Crystals and Rocks under the Microscope

A great deal of information can be obtained about rocks, and about the minerals of which they are composed, by examining the rocks under a microscope. The techniques required for preparing and examining crystals and rocks are very different from those required for biological specimens.

Naturally Occurring Mineral Grains

Collect small samples of sands from different streams, rivers, beaches, dunes and from soft crumbly sandstone. Place a small amount of sand on a moistened slide and examine under the microscope. The *roundness* and *sphericity indices* can be estimated visually using the chart in fig. 73. There are, of course, also variations in gross shape, such as blades, discs or rollers, as opposed to spheroidal grains. Zingg's classification is an attempt to discriminate between them by comparing the ratios of their longest, intermediate and shortest intercepts (fig. 74). The occurrence and depth of hollows (*re-entrants*) in a grain can also be noted. The range of grain sizes can be determined by using a calibrated graticule eyepiece

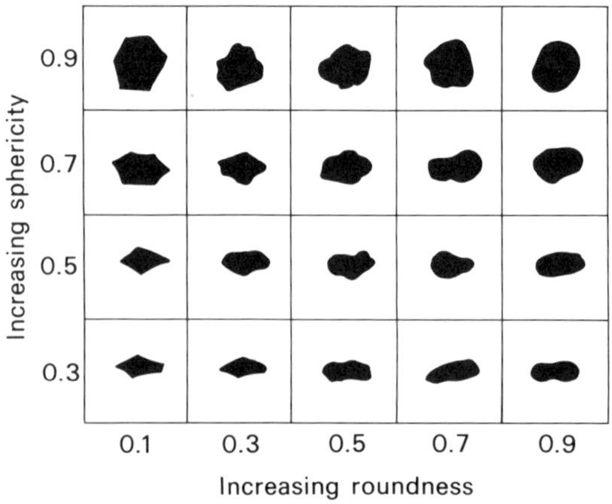

Fig. 73 Chart for visual estimation of roundness and sphericity of grains. Showing increasing sphericity and increasing roundness.

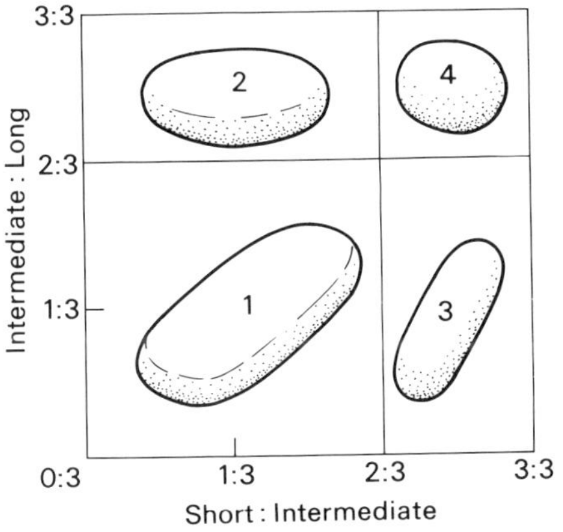

Fig. 74. Visual estimation of grain shape using Zingg's chart. Showing intermediate-long and short-intermediate. (1) Blade. (2) Disc. (3) Roller. (4) Spheroid.

(Chapter 1) or by passing the grains through a series of progressively smaller mesh-sized sieves. These features are largely determined by environmental factors. For example, wind tends to cause high roundness and sphericity indices, whereas most rivers do not. Furthermore, the wind deposits a limited variety of grain sizes, so a narrow range of grain size values might indicate a sand dune origin of the sample.

A permanent preparation may be made by mounting the grains in Canada balsam in the normal manner. All geological mounts are conventionally made in Canada balsam of the natural filtered variety, or a mountant with the same refractive index (1·54), and reference books assume that this is the mountant being used.

Observing Growing Crystals

Crystals can be induced to grow on a microscope slide, and examination will show that the crystal form and colour is characteristic for particular chemical compounds. Common salt (sodium chloride, NaCl) is a good example for this technique. Dissolve most of the salt in warm water leaving a little undissolved. A small drop of this saturated solution is transferred to a slide and thinly spread out. Gently warm the slide from below, by holding it against a lamp or passing it through a bunsen flame, so that some of the water is evaporated, leaving a supersaturated solution out of which crystals will grow. After a while, place the slide on the microscope and observe the edge of the solution, for this is where the crystals suddenly appear.

The crystals of sodium chloride form when the positively charged sodium ions (Na^+) and the negatively charged chloride ions (Cl^-), having been brought close together by the evaporation of the water, form chemi-

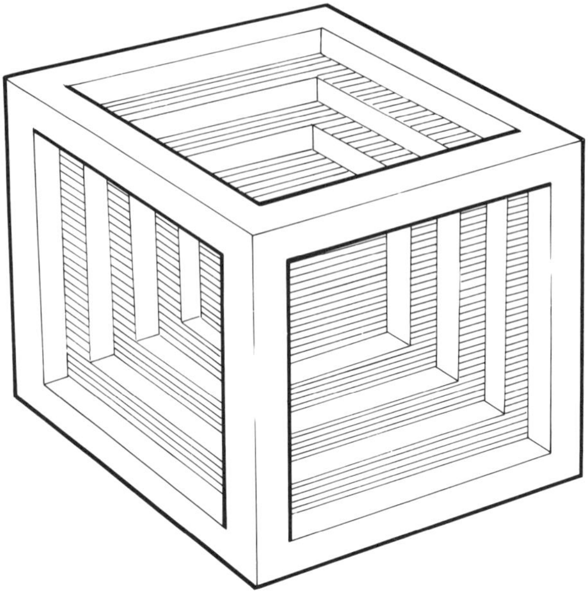

Fig. 75 Hopper-faced sodium chloride crystal, due to faster growth on edges rather than faces, creating steps.

cal bonds which hold them in a regular cubic pattern. In practice, most of the sodium chloride crystals will not be perfect cubes because they will grow faster in some directions than in others, depending upon the rate of supply of ions. Sometimes octahedral and even twinned crystals develop, but these still retain the same cubic symmetry.

A particular curiosity of sodium chloride is that it often grows *hopper-faced* crystals. It seems that the crystal *edges* rather than the crystal *faces* build up excess charge, making them particularly attractive to the sodium and chloride ions, so that the crystal builds up faster at the edges. This faster growth must still obey the principle of one sodium ion surrounded by six chloride ions, and vice versa. As a result, right angled steps appear, creating the hopper faces (figs. 75 and 76).

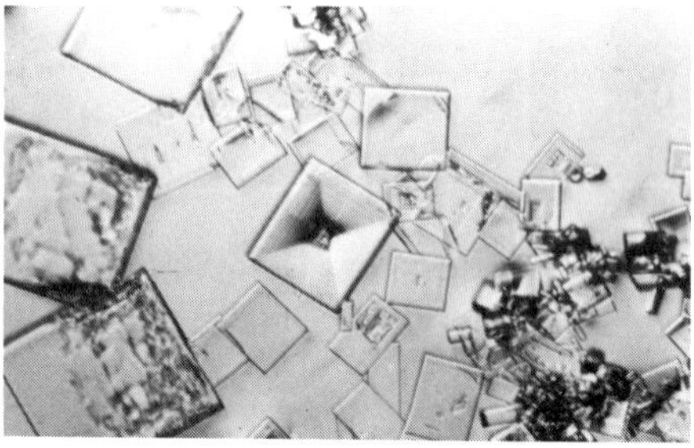

Fig. 76 Hopper-faced sodium chloride crystals (×30).

Other crystalline compounds can be treated in a similar manner, for example:

sodium chloride (Na^+Cl^-) (in pure form rather than as household salt)	
copper sulphate ($Cu^{2+}SO_4^{2-}$)	blue crystals
ferrous sulphate ($Fe^{2+}SO_4^{2-}$)	pale green crystals
magnesium sulphate ($Mg^{2+}SO_4^{2-}$) (or Epsom salts)	arrow shaped crystals
potassium permanganate ($K^+MnO_3^-$)	needle shaped crystals, often extremely long
sucrose ($C_{12}H_{22}O_{11}$)	isolated crystals are difficult to obtain.

Permanent mounts of crystals can be prepared in the usual way, using Canada balsam (natural filtered variety), although there are disadvantages to this method, as some crystals become less visible owing to the loss of relief (see below). For this reason, crystals are normally mounted dry as *micromounts*. This can be achieved by applying a small drop of gum tragacanth to a clean slide and allowing it to dry. Breathe on it to moisten it, and transfer the best of the crystals to it. Lay a ring of cement around the crystals and lower a coverslip.

Boldness of Crystals

Apart from their natural colour, the most prominent feature to the eye, of most crystals under the microscope, is the degree to which they stand out from the surrounding medium. This is known as the *relief* of the crystal

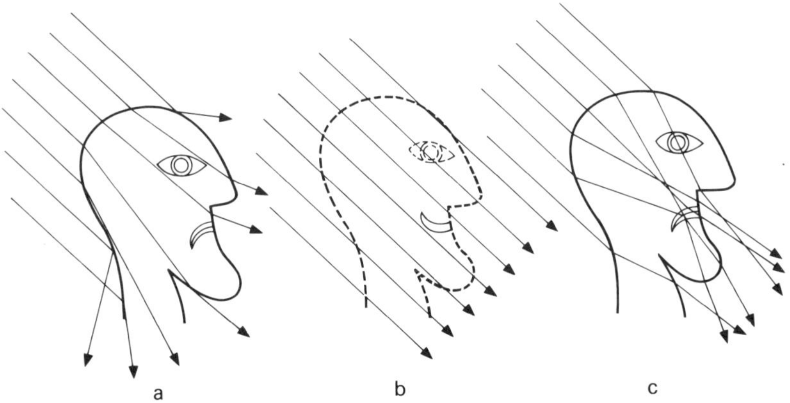

Fig. 77 An explanation of positive and negative relief. (a) Negative relief. Visible transparent man – his optical density is much *less* than that of air, therefore, strong refraction. (b) No relief. Invisible man – his optical density is the same as that of air, therefore, no refraction. (c) Positive relief. Visible transparent man – his optical density is much *more* than that of air, therefore, strong refraction.

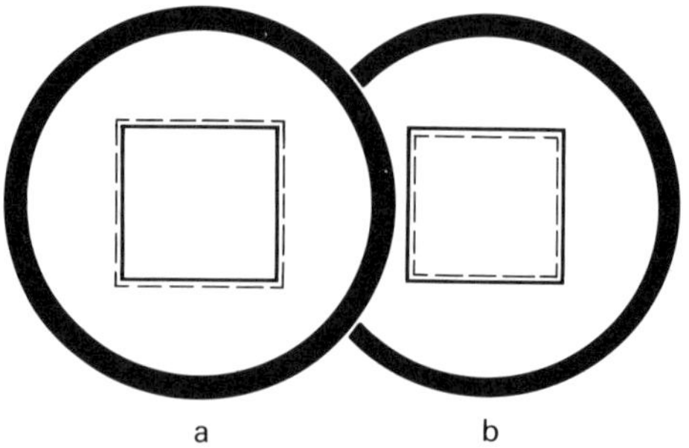

Fig. 78 Becke Line Test. The broken line indicates a bright line seen in the medium with the higher refractive index. (a) The fluorite crystal ($n = 1 \cdot 43$) is mounted in Canada balsam ($n = 1 \cdot 54$). (b) The fluorite crystal ($n = 1 \cdot 43$) is mounted in water ($n = 1 \cdot 33$).

and it can vary from bold (high relief) to faint (low relief). To understand why relief can vary, imagine an invisible man standing in front of you. In order to be invisible it is not enough for him to be transparent. He must also have an optical density (refractive index) identical to that of air. Otherwise the light will be refracted as it enters and then leaves his body (fig. 77). Likewise the relief of a crystal depends upon its refractive index (n), and this can be useful in identifying unknown minerals, since each has its own characteristic refractive index.

The relief of a crystal when viewed through the microscope is determined by the refractive index of the crystals and that of the medium which surrounds it, and can be estimated by subtracting the latter from the former. For example, for the mineral quartz, SiO_2 ($n = 1 \cdot 55$ approx):

quartz in air: 1·55−1·00=0·55 (high positive relief)
quartz in water: 1·55−1·33=0·22 (medium positive relief)
quartz in glycerine jelly: 1·55−1·42=0·13 (low positive relief)
quartz in Canada balsam: 1·55−1·54=0·01 (very low positive relief or zero)
quartz in immersion oil: 1·55−1·52=0·03 (very low positive relief or zero)
quartz in Naphrax: 1·55−1·77=−0·22 (medium negative relief)

The above results can be confirmed by mounting quartz crystals in the mounting media listed and observing them under the microscope. Different values can be obtained for other minerals, such as fluorite ($n = 1 \cdot 434$), which will have medium negative relief in Canada balsam, and olivine ($n = 1 \cdot 64$ to $1 \cdot 88$) which will have high positive relief.

To decide whether the relief of the crystal is positive or negative, carry out the following procedure under high power. Focus on the edge of a crystal in contact with the mounting medium. Increase the contrast by closing the iris diaphragm to the minimum, and then, using the fine focus control, defocus by moving the specimen and objective *apart*. A bright rim of light should be seen either just surrounding or just within the slightly out-of-focus crystal. This rim of light will move into the crystal only if this has a higher refractive index than the mountant (fig. 78). This is known as

Fig. 79 Light is transmitted through this pair of overlapped polars, therefore the optical orientation of the polars must be parallel. Notice the doubling of absorption in the overlap zone.

Fig. 80 No light is transmitted through this pair of overlapped polars (blackness), therefore the optical orientation of the polars must be at right angles to each other. This is known as crossed polars.

the *Becke line test*, and it is used to determine the refractive index of a crystal, if a number of different mounting media with different refractive indices are available. The medium in which the crystal cannot be seen has the same refractive index as the crystal. Standard sets of immersion liquids are available from suppliers, but they require periodic inspection with a refractometer to ensure that the refractive index is stable.

Crystals in Polarised Light

The use of polarised light reveals a great deal of information about crystals, and can help in their identification (plate 4). Any microscope can easily be converted into a *polarising microscope* by using two small sheets of polarising material from a supplier, or from the lenses of a pair of polarising sunglasses. A feature of polarising sheets, readily seen by holding the two pieces up to the light, is that if one is rotated in front of the other, the light intensity through the sheets (*polars*) varies from bright (fig. 79) to almost black (fig. 80). At the black position, they are known as *crossed polars*, and it can be seen that this position, which results from the total absorption of the light, occurs twice in each complete 360° rotation. To modify the microscope, place one of the polars somewhere below the stage, or, if this is not possible, on the stage beneath the specimen. The upper polar (or *analyser*) can be placed over the eyepiece. The effect of the polars is due to organic dyes impregnated into them. These dyes absorb most of the light not vibrating in a particular plane. It is as though each polar has microscopic slits which enable only the light vibrating in that particular plane to pass through. If the second polar is orientated in the same plane as the first, then it poses no extra obstacle but if it is rotated

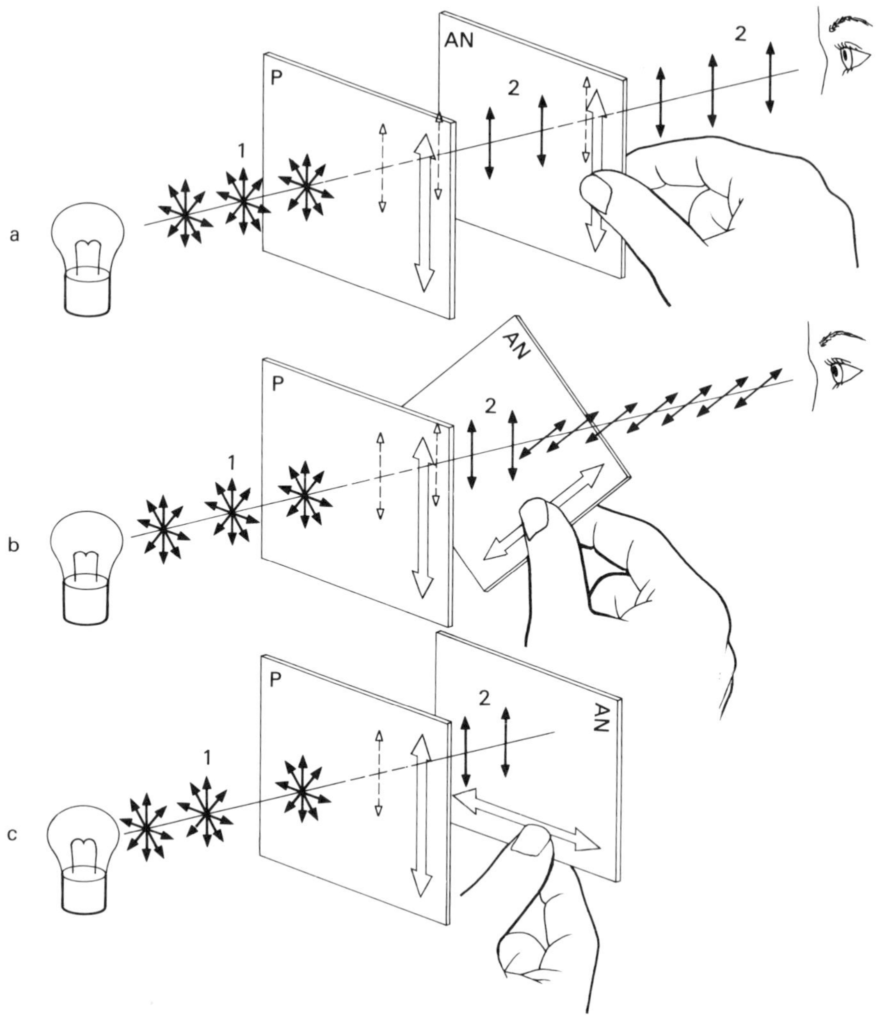

Fig. 81 (a) If polariser (P) and analyser (AN) are uncrossed, plane polarised light reaches eye. (b) If polars are only partly crossed, plane polarised light still reaches the eye, but of reduced intensity. (c) If polars are crossed, no light reaches eye. (1) Unpolarised light. (2) Plane polarised light.

through 90°, then the 'crossed polars' position is reached and light transmission is stopped. Partial crossing of the polars permits passage of a reduced amount of light (fig. 81).

Double Refraction in Crystals

The colour patterns exhibited by crystals under polarised light can be explained by discussing *double refraction*. To see this, purchase a rhombohedral cleavage fragment of clear calcite (Icelandic spar, $CaCO_3$). If the calcite is placed upon a bold black dot on a piece of paper, the dot will appear double, and if the calcite is rotated, one dot appears to be stationary while the other seems to rotate around it (fig. 82). Tilt the calcite so that the separation between the two refracted images diminishes. Eventually a position can be found in which the two images merge back into one another. This merged image may be enhanced by flattening the two corners of the fragment by grinding on a carborundum slip, and then dipping the calcite into cold dilute (about 3%) hydrochloric acid for 10 seconds to make the new faces clear. The image of the dot when the calcite is stood on end will now be more apparent. This position is known as the *optic axis* of the fragment (fig. 83).

Calcite has only one optic axis (optically uniaxial), but many other crystals have a second (optically biaxial).

POLARISATION OF LIGHT CAUSED BY DOUBLE REFRACTION IN A CRYSTAL

The important point about double refraction is that it creates polarised light. This may be confirmed by placing the calcite on a dot as before, but viewing it through a sheet of polarising material. If the polar is rotated, one of the dot images disappears (fig. 84). If the polar is moved through 45°, the first image fades but the second image faintly reappears (fig. 85) and after a further 45° the first dot image vanishes while the second image

Fig. 82 A calcite cleavage fragment creates two dot images when placed upon a single dot. This phenomenon is known as double refraction.

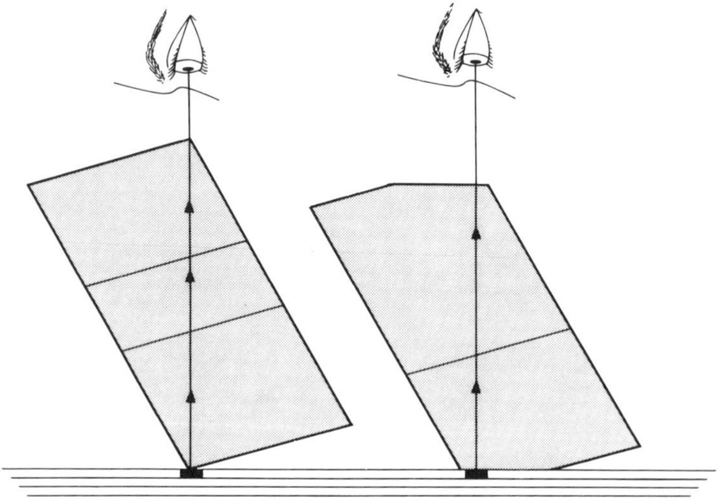

Fig. 83 When only one dot image is visible, the optic axis has been found. The right-hand diagram shows a calcite cleavage fragment cut and etched to clarify this.

becomes strong. Double refraction must therefore be polarising the light into two rays (*ordinary* and *extraordinary rays*) vibrating at right angles to one another (fig. 86). The ordinary ray is defined as the ray which vibrates at right angles to the optic axis of the mineral.

VARYING THE VELOCITY OF LIGHT BY DOUBLE REFRACTION

Place the calcite cleavage fragment upon a dot, and observe that the two dot images appear to be at two different levels in the calcite (*apparent depths*). The light rays forming the images must therefore have different

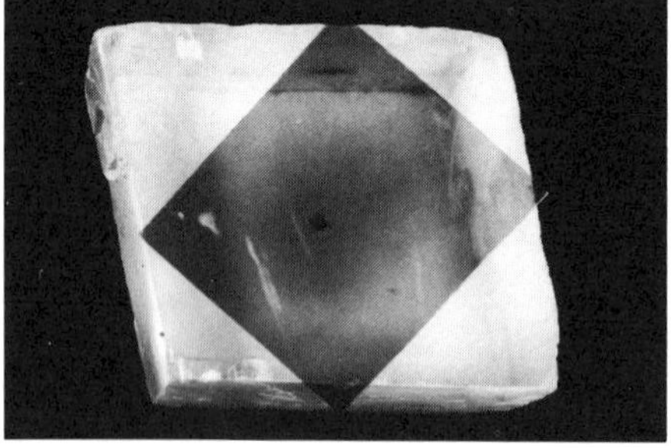

Fig. 84 A polarising sheet causes one dot image to disappear in this position.

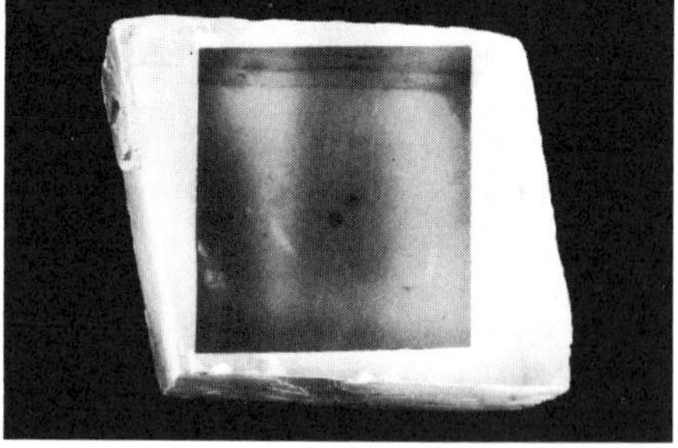

Fig. 85 When the polarising sheet is rotated, the dot image fades and its double appears.

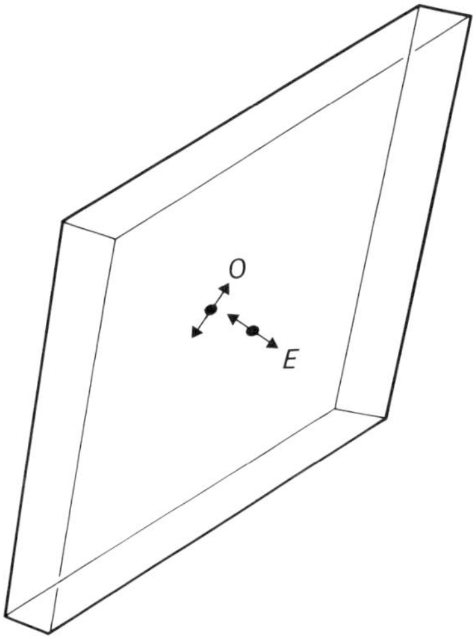

Fig. 86 The rays from the two dot images are plane polarised at right angles to each other. The arrows indicate the plane of vibration of the two rays. *O* is the ordinary ray and *E* the extraordinary ray. The velocity of *E* depends upon the mineral orientation.

velocities, that is, the refractive indices of the ordinary and extraordinary rays are different. This should lead to corresponding changes in the relief of crystals if viewed in plane polarised light. To see this, mount some cleavage fragments of calcite, about 1 mm in size, in Canada balsam, and examine under the microscope with only the lower polar in place. If the slide is rotated (or the polar, if easier), some of the calcite will appear to flash, especially if rotated quickly. This effect, known as *twinkling*, is due to the plane of vibration of the light from the lower polar, lining up first with the plane of the ordinary ray (low relief), then with the extraordinary ray (high relief), and so on, each ray becoming fully visible every 90° (fig. 87).

Closely related to twinkling is the colour change seen in some coloured minerals when rotated with just the lower polar in place. This effect, known as *pleochroism*, is well shown in biotite, when thin sectioned, parallel to its optic axis (fig. 88).

CROSSED POLARISERS AND EXTINCTION

Cross the two polars (place them at right angles to each other) so that the view down the microscope is black. If a clean glass slide, or a slide with a drop of water or Canada balsam, is placed on the microscope stage, nothing will be seen. Similarly, if mounted crystals of sodium chloride or crushed fluorite are placed on the stage, again nothing will be seen. In all of these substances, the internal arrangement is of great uniformity, so there can be no *interference effects*. Such substances are termed *isotropic* in contrast to *anisotropic* substances such as quartz which produce varied *interference colours*.

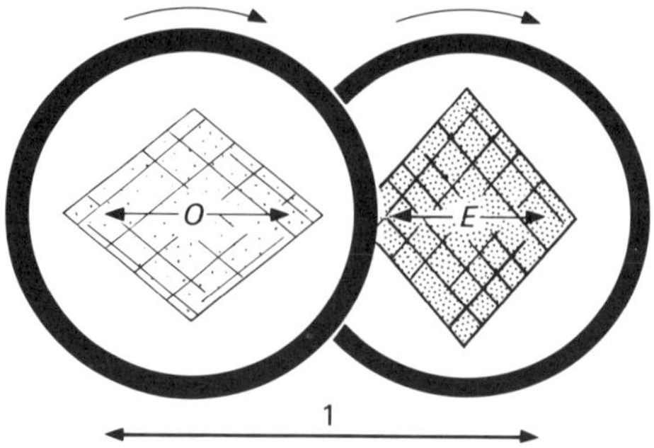

Fig. 87 Rotation of calcite in plane polarised light causes twinkling. In other words, the relief of the mineral, determined by the refractive indices of the ordinary (*O*) and extraordinary (*E*) rays, depends upon which ray reaches the eye. (1) Vibration plane of polarised light. In (*O*) only the ordinary ray is visible ($n = 1{\cdot}49$), therefore, low relief. In (*E*) only the extraordinary ray is visible ($n = 1{\cdot}66$), therefore, high relief.

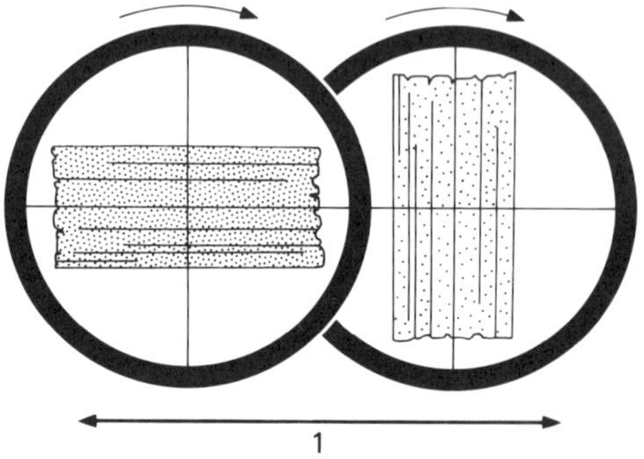

Fig. 88 This is an example of pleochroism, i.e. a colour change in a mineral when rotated under plane polarised light (in this case biotite mica).

No matter how bright the colours transmitted from the mineral to the eye, at some point during rotation the mineral will appear black (the *extinction position*) under crossed polars. There are, in fact, four such extinction positions in each complete rotation of the mineral, one at each 90° of turn (fig. 89).

PETROLOGICAL MICROSCOPES

The relationship of the extinction position of the crystal, to crystal features such as cleavage planes is of great value to geologists. Therefore one of the features of a *petrological microscope* is the fitting of a rotating stage (fig. 90). They also have, in the eyepiece, two crosshairs at right angles to each other, with the east–west crosshair parallel to the plane of polarisation of the lower polar, and the north–south crosshair parallel to the vibration direction of the upper polar. Such an eyepiece or graticule can be purchased from a supplier. Once inserted, it can be oriented by observing the twinkling of calcite in Canada balsam using only the lower polar. When the effect is that shown in fig. 87, then the lower polar is oriented

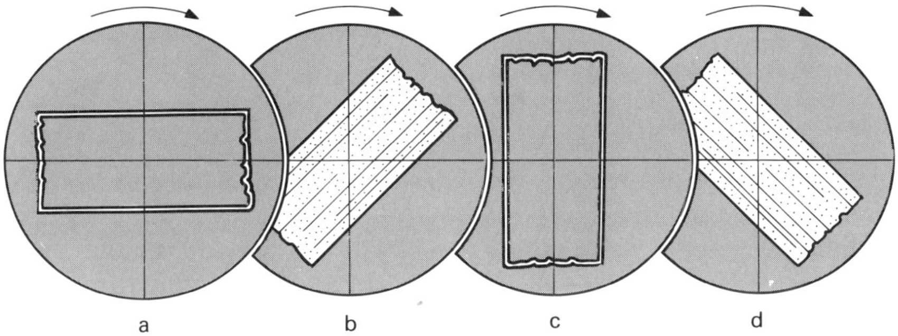

Fig. 89 Under crossed polars, the light from most minerals is extinguished (black) at every 90° of rotation of the specimen (a) and (c). At the 45° positions of the mineral (midway between the extinction positions) the brightest interference colours are displayed (b) and (d). For biotite mica, extinction occurs when its cleavage planes are parallel to the optical orientation of the lower or upper polar; that is straight extinction when parallel to either the north-south or east-west crosshair.

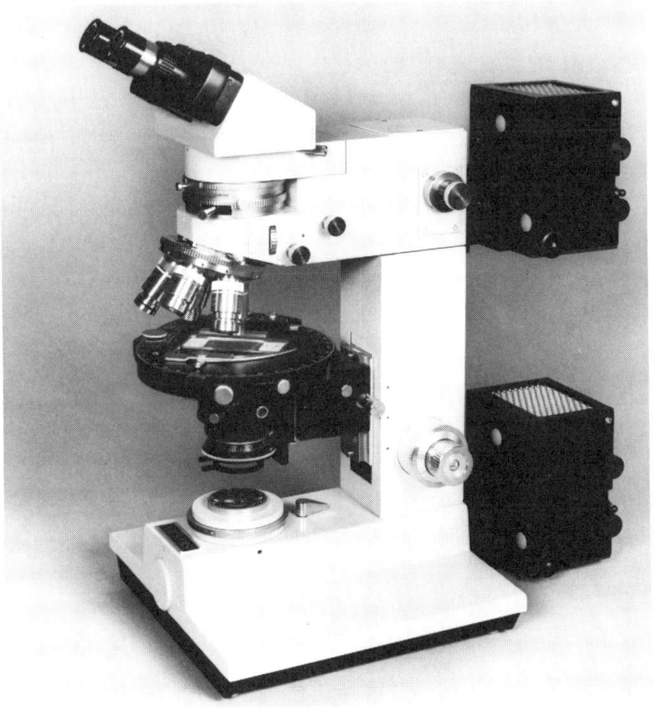

Fig. 90 Vickers M-17 microscope equipped for geological work (petrological microscope). Note the rotating stage.

east–west. The lower polar can be marked so that this orientation may be easily repeated.

EXTINCTION POSITIONS

Minerals are said to have *straight extinction* if blackness occurs when the face or the cleavage of the crystal is parallel to the vibration direction of one of the polars. *Oblique extinction* describes minerals where this relationship does not hold true. A crystal face is due to crystal growth, whereas cleavage is a plane of internal weakness of the crystal along which it can be easily broken. Both faces and cleavage are due to the symmetry of the crystal.

Feldspars show oblique extinction particularly well. Many minerals, such as quartz, lack cleavage, so determination of straight or oblique extinction depends upon finding a good crystal face. Unfortunately quartz in particular often fails to have clear crystal faces under the microscope.

Interference Colours in Minerals

Minerals examined under crossed polars often exhibit interference colours (plate 4). Since the crossed polars should stop all light, the mineral must be altering the polarisation of the light, which is then transmitted through the upper polar. This is achieved, as we have seen, by double refraction in the mineral modifying the incoming light into an ordinary and an extraordinary ray. The colours seen are due to the interference between the ordinary and extraordinary rays, which have different velocities in the crystal and therefore emerge out of phase with a path

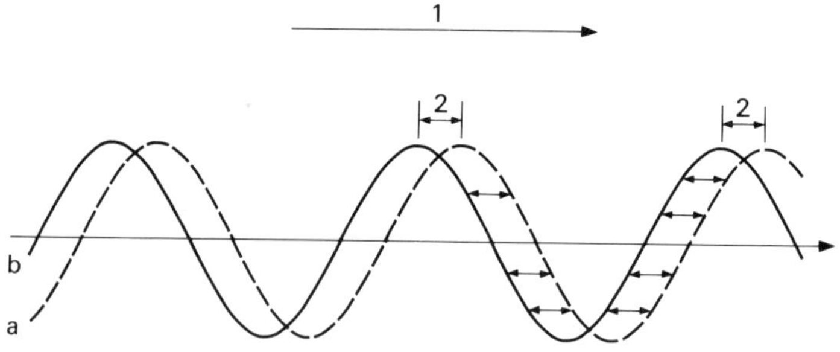

Fig. 91 A particular interference colour is due to a particular path difference between the (a) ordinary and (b) extraordinary rays when they are brought back into the same plane by the analyser for presentation to the eye. One ray lags behind the other (retardation). (1) Light travel. (2) Retardation = path difference in nm.

difference. When the upper polar reorientates the two rays back into the same plane, their path difference allows them to interfere (fig. 91), thereby producing interference colours (plate 4).

The path difference means that either the ordinary or extraordinary ray is *retarded* relative to the other. The amount of retardation (R) depends upon the difference in velocity of the two light rays through the specimen (or the difference in refractive indices n_γ and n_α), and also on the distance over which this difference is maintained (ie the thickness, t, of the specimen). Thus:

$R = t(n_\gamma - n_\alpha)$, measured in nanometres.

Each retardation value produces its own characteristic interference colour.

NEWTON'S SCALE

The interference colours seen under crossed polars belong to *Newton's Scale* and are arranged as *orders*, each rather like a colour spectrum. Low orders are bold, but the high orders become faint, ultimately too faint to see. Newton's Scale can be seen by mounting grains of quartz in Canada balsam and examining them under crossed polars. The colours of Newton's Scale should appear as in plate 7. It can be seen that the colour varies progressively with the thickness of the grain. Working inwards from the edge of the grain (almost no thickness) towards the thicker centre, produces colours which pass through the first order: iron–grey, blue–grey, pale grey, white, yellow, orange, orange–red, red and violet, then through the second order colours and so on, eventually becoming pale and indistinct. The colours of Newton's Scale are given in fig. 92 and plate 7.

A standard Newton's Scale can be made from the transparent variety of gypsum ($CaSO_4 \cdot 2H_2O$) known as selenite. Selenite is soft and has good cleavage, so a cleavage fragment about 2 mm thick can be easily made with a knife. Place the fragment under the microscope to find its extinction position between crossed polars, ensuring that the polars are indeed crossed at exactly 90°. With the gypsum fragment still under the microscope, scratch an east–west line upon the gypsum once you are satisfied that the lower polar is orientated east–west from previous tests with twinkling calcite. Remove the gypsum, and from it cut a rectangle with east–west and north–south sides about 30 mm and 5 mm respectively. Attach it near to the end of a slide using Lakeside cement. Once properly attached, the gypsum can be ground upon a fine abrasive plate to create a gradual wedge. Add Canada balsam and a coverslip. The colour bands of Newton's Scale will be seen when the gypsum wedge is placed in a

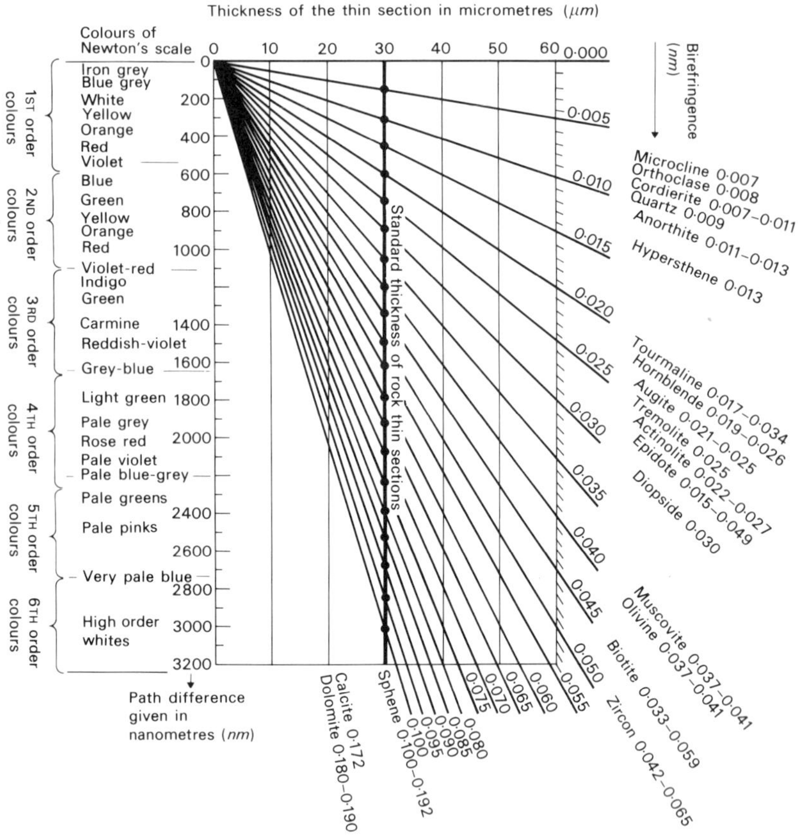

Fig. 92 Michel-Levy Birefringence Chart. The sloping lines relate birefringence values characterising particular minerals, to both thickness of the mineral section and the interference colours displayed. (See plate 7.)

NW–SE or NE–SW orientation on the microscope. Rather more useful but difficult to make, are quartz wedges. These are obtainable from geological suppliers.

MICHEL–LEVY BIREFRINGENCE CHART

A chart can be constructed to show the relationship of Newton's Scale to the retardation, section thickness and *birefringence* (the difference between refractive indices ($n_\gamma - n_\alpha$) within the mineral). This chart is known as the *Michel–Levy birefringence chart*, it can be used to estimate section thickness, and to identify unknown minerals.

For example, a thin section of rock rich in quartz will have quartz grains in various optical orientations. A search may reveal that the highest colour visible under crossed polars is first order pale yellow. From the chart, the retardation creating this colour is estimated to be about 270 nm so:

retardation=270 nm=thickness×birefringence

The maximum birefringence of quartz is given in petrology reference books as 0·009, therefore:

retardation=270 nm=thickness×0·009 nm
therefore, thickness=30,000 nm or 30 μm.

So the chart indicates that the rock section has been made with a thickness of 30 micrometres, which is the standard thickness preferred by geologists.

Suppose that within the same section, grains of an unknown mineral are present, and they are found to have a highest colour of second order yellow/orange. From the chart this will have a retardation of about 900 nm. Since the thickness of the section is already known (30,000 nm), then the birefringence of the unknown mineral can be calculated as 0·030 nm, and this can be looked up in the reference books to identify the mineral. It may, for example, be diopside.

Making Rock Thin Sections

Thin sections of rocks provide information about the constituent minerals, their order of crystallisation and the geological history of the rock. Select a fresh piece of rock, not one that has suffered from weathering, water or lichens. Choose a rock which is not crumbly, since this may disintegrate when thin sectioning is attempted. If any grains work loose when the rock is rubbed with a cloth, then the rock is unsuitable. The best sorts to begin with are the coarser igneous rocks such as granite, dolerite, eucrite, peridotite and diorite which all have grains (crystals) averaging more than 1 mm in size.

Using a geological or lump hammer, smash a fist-sized piece of rock to create fragments, about 30 mm×20 mm and as thin as possible. Several hammer blows may be necessary for this. Wear goggles to protect the eyes from flying rock. Then proceed as follows:

1 Select a thin rock chip and label it with a chinagraph pencil.
2 Take a piece of glass about 2–3 mm thick or thicker and about 30 cm square. Mask the cut edges with tape or grind smooth.
3 Place the sheet of glass flat upon a larger sheet of hardboard and place both upon a firm table to serve as a lapping plate.

4 Put a flat teaspoonful of coarse carborundum abrasive (3F grade) on to the glass. Add a little water to make a paste.
5 Using two fingers of one hand, push the rock chip around in the paste, pressing slightly. Push it over most of the glass to avoid creating bumps in the glass. After a while rinse the rock chip under a tap and dry it with a paper tissue. Hold it near a light to check the uniformity of the ground surface. If it is not uniform repeat the procedure.
6 Repeat for the other side until the fragment is about 3 mm thick.
7 Wash off the coarse carborundum particles from the glass plate, the rock chip, hands, hardboard, etc. Any particles remaining will ruin the section at a later stage.
8 Repeat the grinding procedure for one side only, using a medium carborundum grit (F600).
9 Inspect to ensure that the side is uniformly ground. Now grind the same side again using fine carborundum abrasive (F800).
10 Hold up to a bright light when dry. Look for scratches due to coarse grains or particles of rock dislodged from the chip. If the section is scratched the medium grinding must be repeated until the scratch is removed.
11 Dry the rock slice on a linen towel and place it, prepared side uppermost, upon a thermostatically controlled hotplate at 120 °C.
12 Select a clean, grease-free microscope slide and place upon the hotplate alongside the chip.
13 Using a stick of Lakeside 70C thermoplastic mounting cement, touch one end of the stick upon the rock slice until enough melts to coat the slice completely. Repeat upon the glass slide, but only depositing a drop about the size of the rock slice in the centre of the slide.
14 Wait about 30 seconds to allow time for the cement to penetrate the rock slice. Remove air bubbles with a mounted needle.
15 With forceps lift up the hot rock slice and hinge it, cement side down, on to the middle of the glass slide.
16 Using a mounted needle, press gently in the centre of the slice and move the slice in a small circle. Usually this creates a firmer bond and squeezes out air bubbles.
17 With forceps, quickly remove from the hotplate and place upside down (rock section downwards) on a pad of white paper. Check for air bubbles trapped beneath the section and remove them by moving the glass slide in a small circle over the rock slice, while pressing on it with a pencil. Less than a minute is available before the Lakeside cement solidifies.
18 Let the slide cool, rock section uppermost, preferably on a cold iron slab.

19 Now reduce the thickness of the rock section to about 60 μm using a coarse carborundum abrasive (3F grade) on the glass lapping plate. Avoid creating a wedge. Judge the thickness from the degree of translucency when held up to the light. A wedge will show up paler at one end. Push the section backwards and forwards on the lapping plate.

20 At about 60 μm, rinse hands, sections and glass plate. Now use the F600 medium carborundum, again with to-and-fro motion. Check the thickness frequently with a microscope by wetting the slide and searching for a common mineral such as quartz or feldspar. Stop when their highest colour on Newton's Scale indicates a thickness of 40 μm on the Michel–Levy birefringence chart. At this thickness quartz will display a maximum colour of first order red under crossed polarisers.

21 At 40 μm, again rinse the slide, hands and glass plate. Now use the F800 fine carborundum abrasive to create the required 30 μm standard thickness. Use only the slightest pressure to avoid forming a wedge. Wedging can be checked by focusing at each end of the section. Quartz will now have a maximum colour of first order pale yellow.

22 Scrub the 30 μm thin section with a toothbrush in soapy water to remove loose rock and all traces of abrasive. Check when still moist, using the microscope. Rinse under the tap and dry with a linen cloth.

23 Trim away most of the Lakeside cement from around the sides of the thin section, using a steel straight edge and a single-edged razor blade.

24 Place the trimmed section upon a calibrated hotplate set at 85 °C. Place a clean coverslip upon the hotplate alongside it. Allow the section to heat through. The Lakeside cement will become soft but not liquid.

25 Apply Canada balsam to the top of the rock and to the upper surface of the coverslip. Only use natural filtered Canada balsam.

26 Allow the balsam to cook for about 3 minutes, meanwhile removing any air bubbles with a mounted needle. To ensure that the Canada balsam has cooked, pick up a bead of balsam with a mounted needle and allow it to cool. Now press the bead of balsam on to a hard surface. If the bead breaks up then the balsam has cooked enough. Overcooking causes the balsam to darken by charring, ruining the section.

27 Hinge down the coverslip on to the thin section upon the hotplate, balsam to balsam, using a mounted needle (fig. 29).

28 Still on the 85 °C hotplate, use very gentle pressure from the blunt end of a pencil to squeeze as much surplus Canada balsam as possible from under the coverslip. Anything more than gentle pressure may break the section.

29 Using forceps, remove the slide on to the iron cold plate or to a cool place. Allow at least 10 minutes to cool. Gently wash the slide in cold soapy water and dry with a linen cloth. Apply labels.

MAKING SECTIONS WITH A SAW

A diamond saw can be used to cut the rock into slices about 3 to 5 mm thick. Ideally the saw should be adjustable, with the specimen clamped so that successive parallel cuts can be made. Try to cut nearly, but not completely through each time, so that the slices can be conveniently snapped off by thumb pressure. Trim any obvious irregularities from the rock slice using blunt ended pliers and a rotating coarse grindstone or by cutting on a miniature diamond saw. If the cut surfaces are free from marks caused by blade vibration, proceed directly to step 8 of the method described above.

THE WILKINSON METHOD

This method, devised by D. G. Wilkinson of the Manchester Geological Association, has the advantage of requiring lower temperatures than that given above, but the disadvantage that the mountant used has a refractive index of 1·56, whereas standard geological mounts are made in Canada balsam ($n=1{\cdot}54$). The mountant is transparent super epoxy 'Plastic Padding' resin, supplied as two tubes from hardware shops. When mixed, the contents set to a clear resin. The method is the same as that given above, until step 11, but the hotplate is set at 30 to 40 °C. The hotplate does not need a thermostat, the temperature being less critical. Place the rock slice on the hotplate, smooth side uppermost, and alongside it a glass slide on which equal quantities of the two resin components have previously been mixed. Using forceps, hinge the smooth side of the rock slice down on the resin and apply gentle pressure with a pencil. This must be done rapidly for the resin will be hardening. After an hour on the hotplate the rock slice will be firmly bonded to the slide. Continue as for the above method, until the final mounting of step 24 is reached. Finally place the slide on the hotplate at 30 to 40 °C with a slide alongside for mixing the resin components. Transfer sufficient mixed resin to cover the section, hinge down a warmed coverslip, and heat the slide at 30 to 40 °C for an hour to set the resin. Other epoxy resins might require alteration of this method.

7
Photography through the Microscope

Photography through the microscope (*photomicrography*) is an important method for recording images of the specimens examined, so the images can be seen by others or kept as a permanent record. Many people now own a camera. A 35 mm single lens reflex camera (SLR) with detachable lens is ideal for photomicrography. A fixed lens camera can be used, but photography through the microscope will be more complicated. In both cases, best results are obtained if a trial and error approach is adopted. Purpose built photomicroscopes are available, and most research microscopes have the facilities for the attachment of a camera (fig. 93).

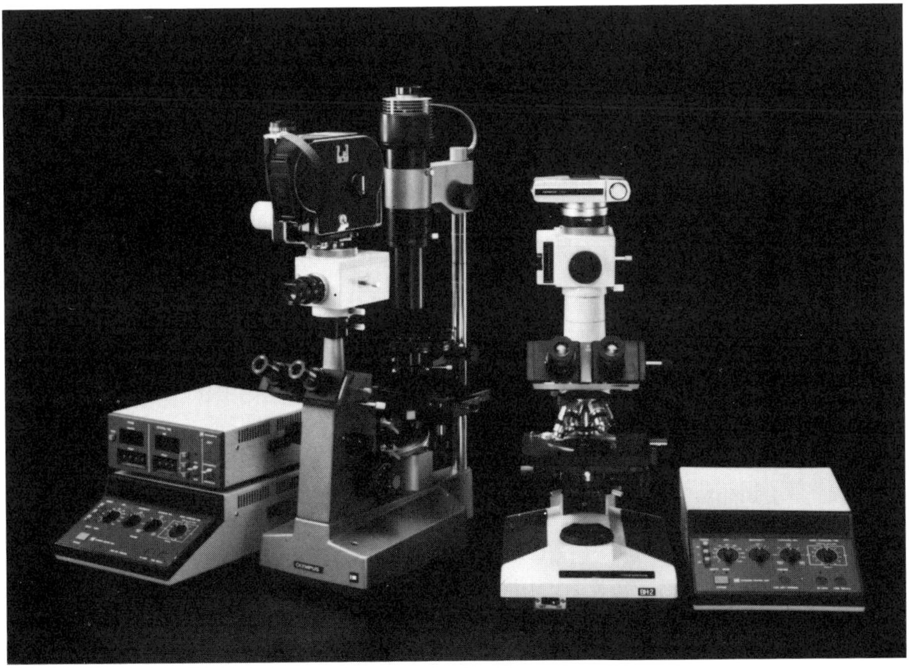

Fig. 93 Modern research microscopes: *right* the Olympus BH 2 complete with fully automatic camera system and *left* Olympus IM Inverted microscope with attached ciné camera.

General Principles of Photomicrography

Before photomicrography is attempted, it is essential to be familiar with the camera and the accessories that are available for it. A knowledge of photography is useful and the reader is referred to books in the Bibliography. Automatic cameras are not generally suitable for photomicrography, unless they have full manual override and brief time (B) settings. In photomicrography, the camera is placed over the eyepiece of the microscope, with suitable masking to stop extraneous light, and the image is projected on to the film. A bright light source is essential so that accurate focusing is possible and short exposures can then be used to minimise external vibration, which can degrade the quality of the photographic image. If the microscope has a mirror, a slide projector can be used as the illumination source, as these generally have very bright lamps. Long exposures are usually required, so it is essential to use a cable release to trip the camera shutter.

The quality of the final photograph depends largely upon the quality of the microscope optical system and how well the microscope is adjusted. Because of the long exposures required in photomicrography, it is best to start with prepared slides or stationary objects. Photomicrography of moving objects can be attempted later using a very bright lamp or flash.

Initially, it is advisable to photograph specimens magnified about 100× (×10 objective with ×10 eyepiece) as higher magnifications need much longer exposures, and the image, being less bright, is harder to focus. A camera with a range of slow shutter speeds and brief time setting is required. All the details of the photographs taken should be noted, so that any faults can be rectified subsequently.

PHOTOMICROGRAPHY USING AN SLR CAMERA WITH INTERCHANGEABLE LENS

Modern single lens reflex cameras are highly versatile. Many have a through-the-lens light metering system, and some have interchangeable viewfinders (pentaprism, waist level finder, critical focus finder, and so on) and/or focusing screens (installed focusing screens normally incorporate various combinations of ground glass, microprism and split image rangefinder). People seriously interested in photomicrography may prefer to buy a clear focusing screen, which can make focusing the microscope image on to the camera film plane easier. Some photographic manufacturers produce adaptors to fix an SLR to a microscope eyepiece tube, but the camera can be fitted just as easily by using suitable diameter tubing (cardboard or plastic). Purpose built adaptors rigidly attach the camera to the top of the microscope tube (note that the eyepiece must be placed in

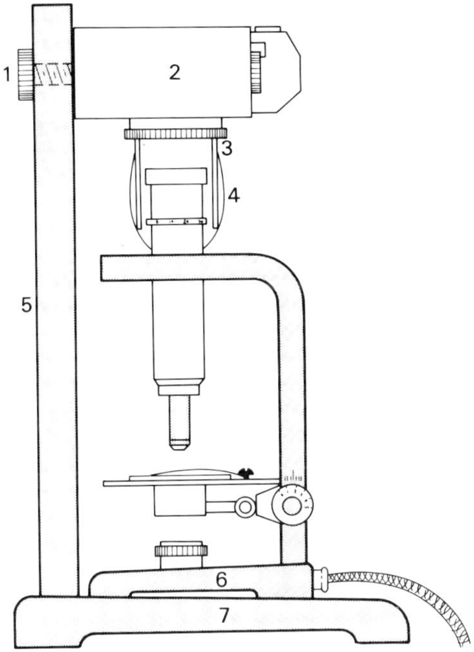

Fig. 94 Stand used to support camera above microscope. Note position of elasticated collar to prevent extraneous light from reaching the film. (1) Bolt. (2) Camera. (3) Tube. (4) Elasticated collar. (5) Camera support. (6) Microscope. (7) Base.

the adaptor). This arrangement is very convenient but can induce vibration when the shutter is fired (an SLR mirror generally induces some vibration, but on some cameras the mirror can be locked up, prior to tripping the shutter). In addition, the microscope focusing mechanism, if on the body, needs to be stiff enough not to drop under the weight of the camera. The microscope design of Chapter 2 makes a good photomicrographic stand, as the tube with camera attached can be firmly clamped into position, and the fine focusing can be accomplished independently by moving the specimen stage (figs. 22 and 95). If possible, a tripod or suitably constructed stand should be used to hold the camera independently above the microscope (fig. 94). Tubes of various lengths (or photographic bellows if available) can be interposed between the camera and microscope, so that little or no 'cut off' is apparent at the edges of the field of view. A 70 mm tube, measured from the top of the microscope eyepiece to the camera body, has been used with good results. The inside of the tube should be painted matt black to minimise internal reflection which can degrade the quality of the photographic image. This tube can be attached to the camera body in various ways. One way is to use a suitable T2 mount. These normally allow lenses to be retained when swapping to a

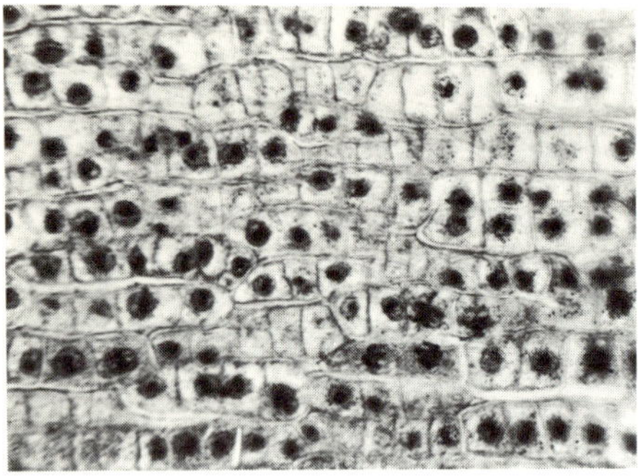

Fig. 95 Photomicrograph of plant cells taken through the more complex version of the home-made microscope (with condenser and mechanical stage) (×170).

camera with a different lens mounting. Purchase one from a photographic dealer, for the appropriate make of camera, and push, glue or tape the tube to this mount. Attach the tube to the camera, and mount the camera over the microscope. It is good practice to use a small spirit level to ensure that the camera is horizontal and at right angles to the microscope. A light trap must be fitted between the camera and microscope. A concentric tube light trap or an elasticated sleeve of thick light-proof material is easy to construct.

The microscope must, of course, be properly adjusted before attempting photomicrography (see Chapter 1), but once this has been done, and the camera suitably mounted over the microscope, photographs can be taken. The specimen on the microscope stage must be accurately focused, so that it is sharp at the film plane of the camera. If the microscope illumination is bright then this is fairly easy. It can be made even easier, if a clear glass focusing screen is fitted to the camera. However, with practice it is perfectly possible to achieve good critical focus with an ordinary microscope and a camera with a standard focusing screen.

PHOTOMICROGRAPHY USING A FIXED LENS CAMERA

The basic requirement is to open up the lens to its maximum aperture or *f* stop (generally *f*2.8 or *f*3.5 with fixed lens 35 mm cameras). A smaller *f* stop will reduce the field visible at the film plane, which may not cover the entire film area anyway (fig. 99). Set the focus at infinity (∞ symbol). The camera should be mounted above the microscope eyepiece so that the front surface of the camera lens is at the eyepoint of the microscope eyepiece. The *eyepoint* can be determined by holding a piece of white paper over the eyepiece and gradually raising and lowering it; the bright circle of light which appears on or through the paper, becomes smaller, then increases again. The eyepoint is the position at which the spot of light is smallest. This eyepoint will have to be determined again if a different eyepiece is used. The camera lens must be levelled and placed symmetrically over the eyepiece, and a cable release used to trip the shutter.

Focusing the microscope, to achieve a sharp image on the film, can best be performed by focusing the microscope with 'relaxed eyes'. It may be possible to relax the eyes by staring at a very distant object just prior to focusing the microscope. If spectacles are normally worn they should be kept on for this. If a poorly focused image results then, instead of this method, use a telescope or one side of a pair of binoculars, focused at infinity, positioned over the microscope eyepoint. The microscope should then be focused while viewing the image through the telescope. Whatever method is used, prior to taking photographs the camera should be placed over the microscope with the front of the lens at the eyepoint. Extraneous light can be eliminated by masking the space between the camera lens and the microscope eyepiece.

Choice of Film

This is largely a matter of personal preference but some guidelines may be useful. Black and white film is often used for photomicrography of scientific illustrations because few printers can justify the high cost of colour reproduction. Black and white films should be of the panchromatic variety (film sensitive to all visible colours) since many specimens will be stained. A fine grain, high contrast film (such as Ilford Pan F or Kodak Panatomic X) is preferable. However, if the exposure time is unacceptably long or external vibration is a problem, a faster film may be better (Ilford FP4 or Kodak Plus X). Exposure latitude is the range of exposure allowable from underexposure to overexposure while still producing a printable negative. Generally, black and white film has a wide exposure latitude, biased towards overexposure tolerance. Also, high contrast film generally provides the least exposure latitude, and low contrast film the most. Fine grain film gives good image quality. The photographs reproduced here were taken on either Ilford FP4 or Pan F material, developed in Paterson Acutol (FX14) developer. This type of film is ideal for experimenting in photomicrography and results can be obtained quickly if home processing equipment is available.

The light from electric lamps always appears white to the human eye. However, unlike the human eye, colour film is unable to compensate for small changes in the colour quality of light. For this reason, colour films are normally formulated for exposure by light of a specific quality, such as daylight or tungsten illumination. These values are usually marked on the film carton as *colour temperature* in degrees Kelvin (K). Kelvin values indicate the colour temperature of the light source to be used in exposing the film. A low colour temperature film (3200K) needs to be exposed by a tungsten filament light source (tungsten film), whereas a high colour temperature film (5500K) needs to be exposed by daylight or electronic flash (daylight film). In colour photomicrography the film used will depend upon a number of factors.

If the microscope has a substage mirror, the specimen can be illuminated with daylight and a daylight colour film used. Daylight colour film can be used with microscopes with built in illumination sources if a *colour compensation filter* is used to modify the quality of light reaching the film. A tungsten light source needs to be used in conjunction with a blue filter (filter No. 80A or filter B12 in the DecaMired system), or else the transparencies produced on colour daylight film will have an overall yellow coloration. Details of the required types of blue filter can be found in Kodak's excellent booklet of photomicrography (see Bibliography).

For general photomicrography using a microscope with built in illumination, a tungsten type colour film is recommended, and these can generally be purchased from larger photographic dealers, who deal with professional photographers. The colour films used for this book were either Kodak Ektachrome 160 tungsten film or Ektachrome 50 tungsten film. These transparency films generally give brilliant colours.

Colour film can be used creatively with special illumination of the specimen, such as Rheinberg illumination and polarised light, see Chapter 6 and 8, or can record the colours of stained sections. True colour rendition, with the types of equipment available to most people outside colleges and hospitals, will not be perfect, but can be most attractive (achromat objective lenses are only corrected for two colours so the transparencies may appear to be slightly unsharp). Colour film has another disadvantage in that some time can elapse while the film is being processed, and thus mistakes and their correction can be a tedious and expensive business, whereas home processing of black and white film quickly shows any mistakes. By using a short length of black and white film, and a trial and error approach, many mistakes can be corrected quickly and easily without the frustration of waiting for professional processing.

Colour negative film (print film) is rarely used in photomicrography as the processing laboratory will have difficulty with colour rendering of unfamiliar objects. Prints can be obtained from transparencies anyway.

The Use of Filters

Colour filters are often used with black and white film to enhance contrast.

Colour stains may appear to have equal brightness on black and white film and may thus be recorded as equal grey tones. Employing a suitable filter, to absorb one colour more than another, will render that colour darker grey and increased contrast will result. Colour filters used with colour film can produce coloured backgrounds to certain specimens and this can make the transparency more aesthetically appealing. The use of a medium density green filter, fitted in the filter holder of a microscope (or beneath the stage of the microscope produced here) can increase the contrast with the common stain combination of haematoxylin and eosin.

Determination of Exposure

Exposure depends upon film speed (ASA or Din setting), light intensity (image brightness), and exposure time (shutter speed). Film speed should be between 50 and 125ASA for black and white film. Image brightness is largely determined by the microscope lamp used, condenser setting (if a condenser is fitted), magnification used, filter factor, and length of extension tube used to fit camera above the microscope eyepiece. Many of these will be constant, if the same equipment is always used, so that alteration of exposure time alone is required to give the correct exposure on the film. The correct exposure time can be determined either by using an *exposure test* or by using a sensitive exposure meter. An exposure test is carried out by setting up the microscope for use, including any filters (coloured filters alter exposure in photomicrography just as they do when fitted to the camera in normal use) and exposing several frames, each for a different length of time. Do not alter the microscope illumination during this test, only the shutter speed of the camera. The exposed film should then be processed and individual negative frames examined to determine which one received the correct exposure. Once determined, one can take photographs with some confidence, but it is still desirable to 'bracket' exposures, that is, also take photographs at the shutter speed on either side of the predetermined exposure. For example, if your test has indicated that a one second exposure is correct, then take photographs at $\frac{1}{2}$ second, 1 second and 2 seconds. This may seem very wasteful of film, but is necessary to compensate for slight differences in the light transmission of specimens, or when re-examination of the specimen is not possible.

Cameras with through-the-lens exposure metering systems can simplify the determination of exposure through the microscope, but in some cases the metering system and the timed shutter speeds available are inadequate for photomicrography. If the image is bright enough and a fast film is used, a short exposure will be indicated by the camera, but this is not always so. Long exposures need a cable release and should be

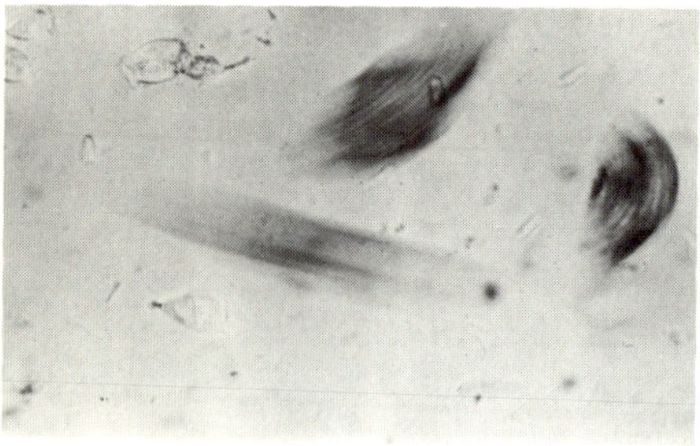

Fig. 96 Blurring of the fast moving ciliate *Blepharisma* (×150).

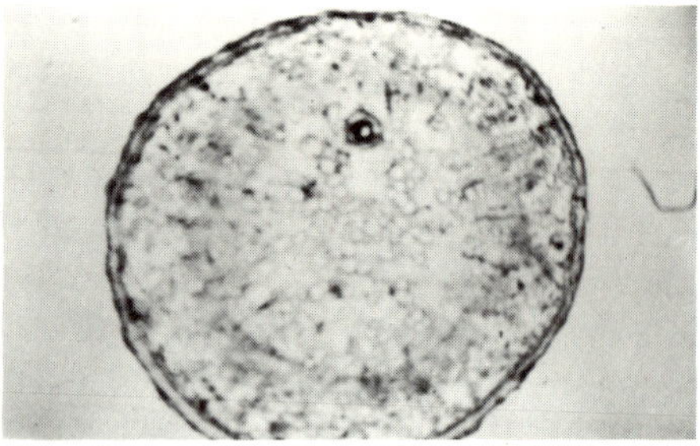

Fig. 97 Out of focus TS of *Rheum* root (×15).

accurately timed. If a stopwatch is not available, count thousands, 'one thousand, two thousand' (you have counted two seconds). By counting in this way an exposure can be timed reasonably accurately.

The camera eyepiece should be masked while determining exposure with the camera meter, as extraneous light can give a false reading (a viewfinder eyecup may be useful here, and can be purchased for most cameras). Furthermore, exposure meters may be of two types, the averaging and the centre weighted type. These may give different exposure readings when used, but generally the centre weighted type is preferable, although, some degree of exposure compensation may still be necessary. Some automatic exposure cameras are fitted with a dial for over or under exposure compensation. With some specimens, one or two stops over exposure will give a correctly exposed negative of the specimen – this is because photography through the microscope is similar to backlighting in general photography (the object is under exposed) where increased exposure is necessary to give a correctly exposed subject on the negative. If an exposure compensation device is not available then halve or quarter the indicated film speed on the camera, and this should give the correct amount of over exposure.

Some Faults in Photomicrography

a) Blurring (fig. 96). This is caused by vibration, or a moving specimen. This can be rectified by isolating the cause of vibration, or (sometimes) by using a faster film, brighter lamp or electronic flash.

b) Out of focus image (fig. 97). This is caused by a poorly focused microscope, or by the camera weight altering the focus during exposure. This can be rectified by carefully checking for critical focus before tripping the camera's shutter or by tightening the focusing mechanism so it does not move under the weight of the camera.

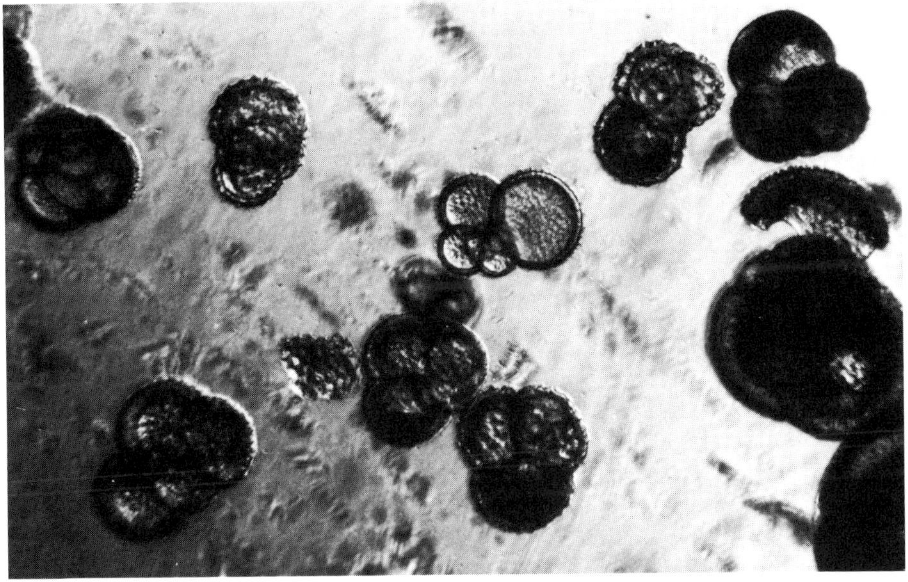

Fig. 98 Photomicrograph of foraminiferal ooze showing poorly-centred illumination. The intensity of illumination varies across the photograph (×100).

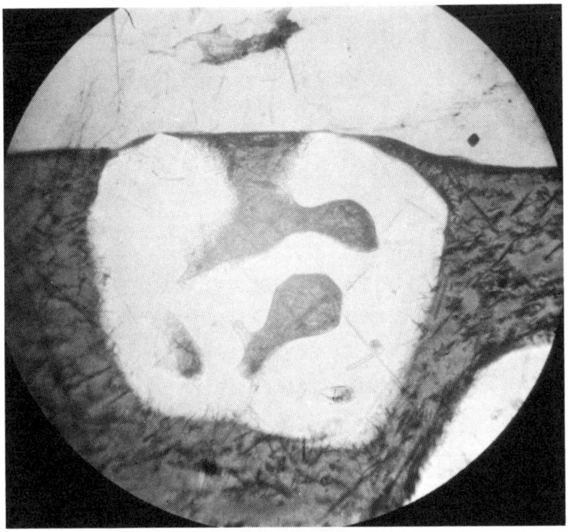

Fig. 99 Photomicrograph of a thin section of pitchstone showing a corroded augite crystal in a glassy matrix. Photograph taken with a fixed lens camera on the simpler version of the home-made microscope. Note the restricted field of view (×30).

c) Uneven illumination (fig. 98). One side of the photomicrograph is illuminated more than the other. This is caused by poorly centred illumination. Check that the lamp and condenser are correctly centred.

d) A restricted field of view (fig. 99). This can be rectified by using a greater separation of camera to microscope or by opening up the camera lens, if of the fixed lens type.

Magnification Calculation

For a fixed lens camera, use the formula:

$$\text{film magnification} = \text{microscope magnification} \times \frac{\text{focal length of camera lens (mm)}}{250}$$

This will be further magnified if the negative is photographically enlarged.

For an SLR camera use:

$$\text{film magnification (mag.)} = \underset{(\text{objective} \times \text{eyepiece})}{\text{microscope mag.}} \times \frac{\text{separation between eyepoint and film plane in mm}}{250}$$

Exposure Calculation

Use an initial test to give a standard exposure. If the objective lens (obj) is changed then the new exposure can be determined by the following formula:

$$\text{new exposure} = \underset{(\text{seconds})}{\text{standard exposure}} \times \left(\frac{\text{standard obj NA}}{\text{new obj NA}}\right)^2 \times \left(\frac{\text{new mag}}{\text{standard mag}}\right)^2$$

This formula only holds true if the distance from film plane to eyepoint remains constant.

8
Some Special Methods in Microscopy

The methods described so far have mostly involved viewing the specimen in conventional bright field transmitted illumination, where the image is formed by passing light through the specimen, which is therefore seen against a bright background. Many cells and even whole aquatic organisms are, however, difficult to see under these conditions, and staining is not always desirable or possible. In these cases other techniques are available, these usually require the modification of the microscope's optical system, rather than modification of the specimen. Some of these techniques are described in this chapter. Of these techniques, some are easily and cheaply attainable with any light microscope, whereas others require expensive additions or alterations to the microscope, and are thus only practicable for research use.

Dark Field Illumination

With *dark field* illumination the specimen is seen brightly illuminated against a dark background (fig. 100), unlike *bright field* illumination, which has been used in previous chapters. Many types of specimen can be examined by dark field illumination, but protozoa, planktonic animals, diatoms, foraminifera and mineral grains are particularly good.

In its simplest form, this method involves placing a suitably sized dark patch beneath the microscope condenser, to impede the light which would normally travel straight into the objective (fig. 101). Peripheral light rays

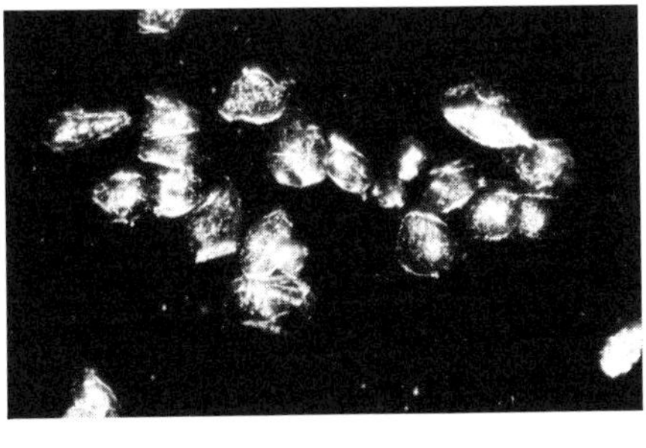

Fig. 100 Cheek epithelial cells brightly highlighted in dark field illumination (×70).

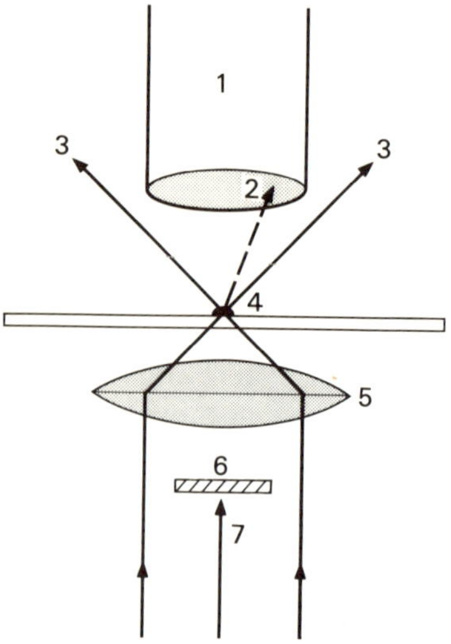

Fig. 101 Diagram to show the action of the patch stop in producing dark field illumination. Note that central rays of light (7) are inhibited by the patch stop (6) from entering the objective (1) and that peripheral rays (3) also miss the objective. Only the light refracted (2) by the specimen (4) enters the objective. (5) Condenser.

travel past the edges of the patch stop and are converged as a wide cone of light towards the specimen. The rays of light in this cone would not normally enter the objective lens aperture, but they will if refracted by the specimen. The specimen thus appears brightly illuminated against a dark background.

A simple *patch stop* is all that is required for use with low magnifications, up to a ×10 objective (fig. 102). A patch stop can be cut out of cardboard and should fit into the filter holder of the substage condenser. The size of the central disc of the patch stop can be determined by removing the eyepiece with the specimen in focus and, looking at the back focal plane of the objective, opening and closing the iris diaphragm until the edge of the aperture is just visible at margins of the lens. The diameter of the aperture in the iris diaphragm is that of the required patch stop for the objective in use. Some trial and error may be required, but all direct rays of light from the illumination source should be blocked out.

To set up dark field illumination using a patch stop, bring the specimen into focus using bright field illumination. Then place the patch stop into the filter holder and adjust the position of the condenser to achieve the desired highlighted specimen against a dark background. A condenser is, of course, essential for this type of illumination and the numerical aperture of this must be greater than the numerical aperture of the objective being used. The iris diaphragm should remain fully open.

For high magnification, high resolution dark field microscopy, special dark field condensers and very bright light sources are required. An example of the use of this would be in the examination of bacterial flagella.

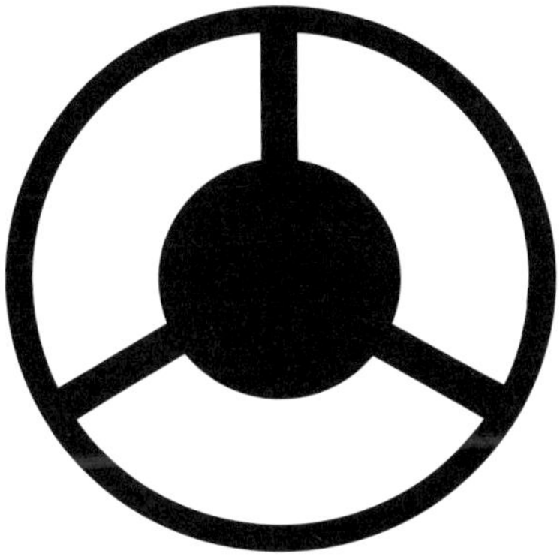

Fig. 102 A patch stop for producing low magnification dark field illumination.

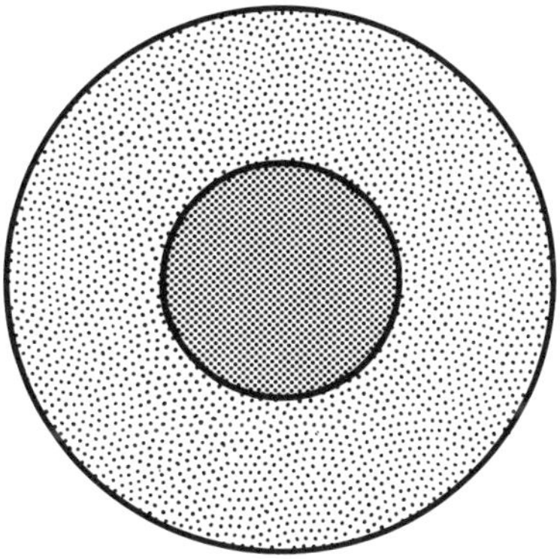

Fig. 103 A Rheinberg disc consisting of a central dark colour and an annulus of a second lighter colour.

These are normally below the limit of resolution of the light microscope, but in dark field microscopy, the light diffracted (*diffraction* is a phenomenon where light spreads out when passing through apertures or around obstacles) by the flagella produces an image which is easily seen against the black background.

Rheinberg Illumination

This is a modified form of dark field illumination where, instead of the specimen appearing white against a black background, it can be of any colour set against a background of a different colour. Instead of the patch stop used for low magnification dark field illumination, Rheinberg illumination uses a celluloid disc with a coloured centre and a differently coloured peripheral strip (annulus) surrounding it (fig. 103). Light of one colour (the background colour) passes through the central part of the disc into the objective, whereas peripheral coloured rays (the specimen colour) illuminate the specimen. For example, if the central disc is red and the outer circle yellow, then the specimen will appear yellow against a red background (plate 5). These discs are called *Rheinberg discs*, and they can be made up from coloured celluloid bought from an artist's supply shop. The central coloured material should be of a darker colour than that of the surrounding annulus.

A more complicated Rheinberg disc uses three colours. The central disc is made of one colour, while the annulus is composed of four segments made of two different colours, placed diagonally opposite one another.

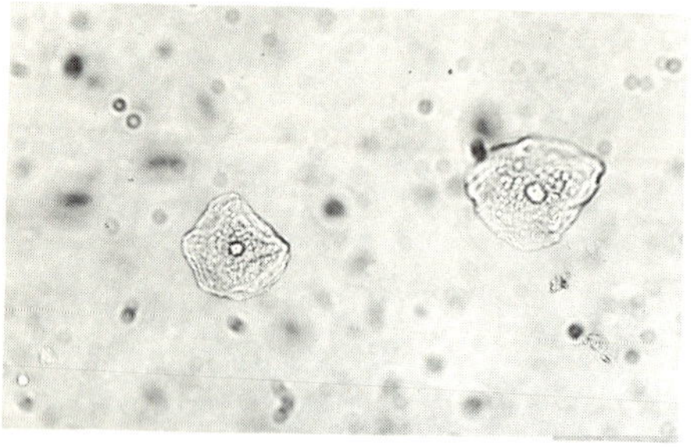

Fig. 104 Cheek epithelial cells observed with the condenser stopped-down to accentuate contrast within the cells. Compare with figs. 32 and 33 (×150).

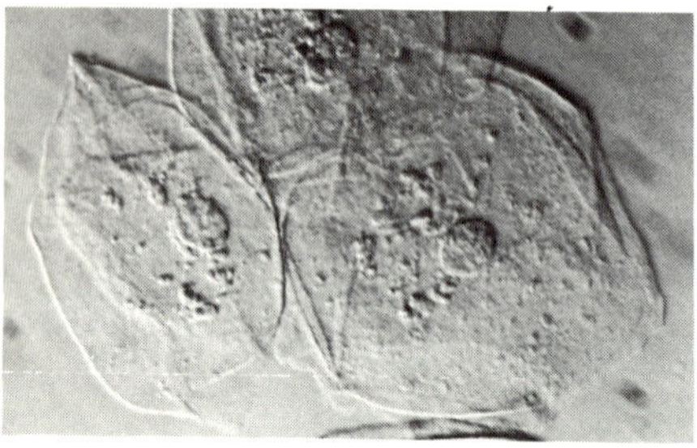

Fig. 105 Cheek epithelial cells in oblique illumination displaying three-dimensional appearance. Compare with figs. 32 and 104 (×400).

Fig. 106 The ciliate *Colpidium* in oblique illumination. The internal vacuoles are particularly distinctive in this illumination (×170).

When material is examined with this type of Rheinberg disc in place, a structure running in one direction appears in a different colour from one running at right angles to it, set against a background of the third colour. For example, correctly orientated woven material can be examined in this way. Rheinberg illumination is particularly striking when used to illuminate unstained specimens for colour photomicrography (plate 5).

Stopped-down Contrast

Very little detail can normally be seen in unstained living cells under bright field illumination, because of the lack of contrast between the specimen and the surrounding medium. However, these specimens can be examined using a microscope fitted with a substage condenser by adjusting the substage iris diaphragm.

The specimen slide should be placed on the microscope stage and brought into focus. If the specimens are very low in contrast, then a mark made on the coverslip during mounting will aid focusing. To increase specimen contrast, close down the substage iris diaphragm considerably,

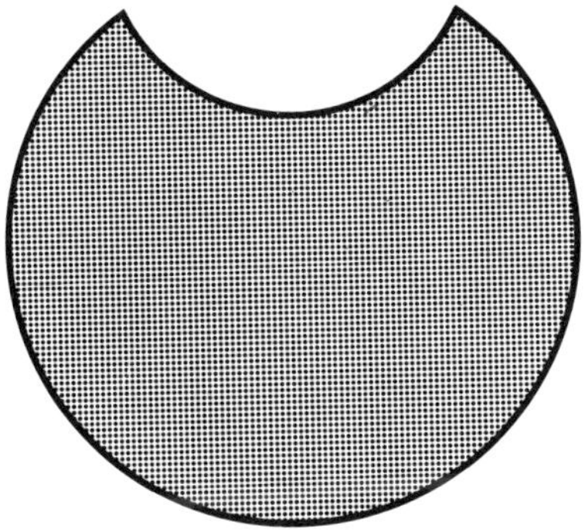

Fig. 107 Stop with peripheral segment removed for use in production of oblique illumination contrast.

and readjust the fine focus, so that the specimen is slightly out of focus, which accentuates the contrast (fig. 104). Closing the diaphragm will result in some loss of resolution but will greatly increase the contrast of transparent specimens.

Some practice may be required to produce the optimum amount of contrast in a given specimen by this method. Because the adjustment of the iris diaphragm makes the intensity of illumination low, dust particles in the optical components, especially the eyepiece, become very noticeable, and may require removal with a blower brush or lens tissue.

Oblique Illumination

By illuminating the specimen obliquely one can produce a relief effect which accentuates any phase gradients within a transparent specimen (figs. 105 and 106). This relief effect can be achieved by moving the condenser laterally.

Condensers are available which allow some of their optics to be slid sideways, normally for dark field work. Trial and error is required to produce optimum results. This technique can be used with objectives up to ×40 magnification.

This effect can also be produced with most of those microscopes fitted with a normal non-sliding condenser, by placing an opaque disc, which has been modified by removing a peripheral segment of an appropriate size, into the condenser filter holder. Such a disc with a peripheral segment removed is illustrated in fig. 107. The size of the removed segment can be found by trial and error. Each of the dry objectives fitted to the microscope may need its own disc, which can be appropriately identified. Filter holders which swing into the optical path of the condenser are useful for this technique, as a disc can be accurately positioned to produce the desired relief effect.

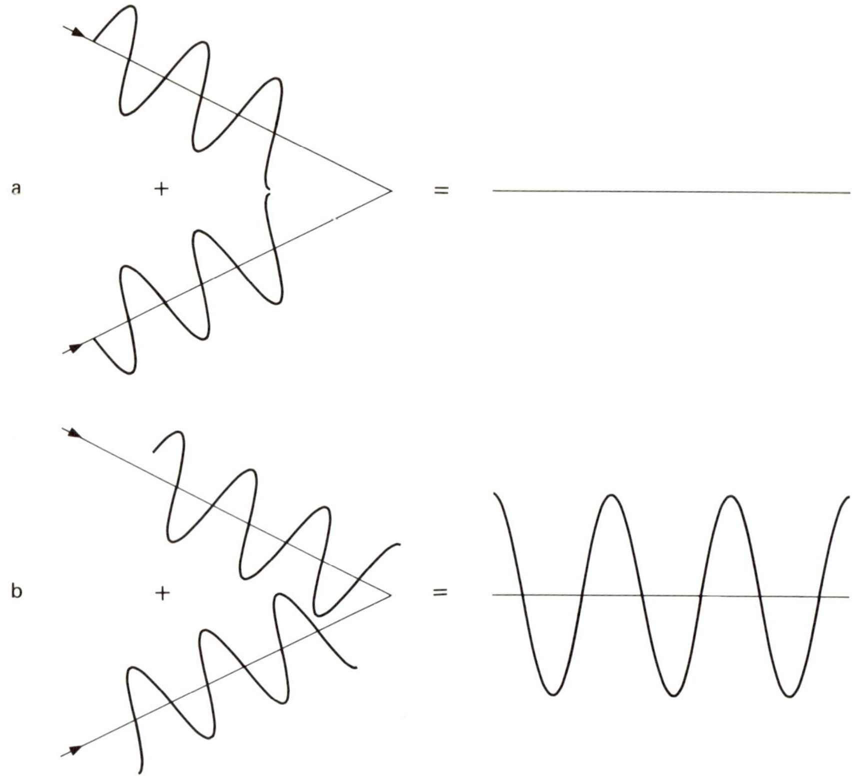

Fig. 108 Diagrammatic representation of the interference of waves. (a) When two waves of equal wavelength meet half a wavelength out of phase they produce destructive interference with one another, so that no wave train exists (darkness). (b) If the same two waves meet in phase, they combine additively to produce a wave of increased amplitude (brightness).

Phase Contrast

Little contrast is apparent in living cells, despite the differences in refractive indices of the specimen and surrounding medium. This is because the eye is not sensitive to phase changes between light passing through and light passing around the specimen. However, if phase changes are converted into a change of amplitude (intensity), then they can be perceived.

If two identical light waves meet, such that the peak of one wave meets the peak of the other, they are said to be *in phase* and will add together to produce a new wave with a greater amplitude than either of the original waves (fig. 108b). If, however, the two waves meet so that the peak of one corresponds with the valley of the other (half a wavelength *out of phase*), then they will *interfere* with and destroy one another, so that no wave now exists (fig. 108a). This is called *destructive interference* of waves. All intermediate interactions are possible. When light passes through a specimen it will be slowed down (retarded) or quickened (advanced) depending upon the refractive index of that specimen, and the same will occur to light passing through the mounting medium, but to a different extent if the refractive index is different from that of the specimen. If light passing through the specimen is combined with light passing through the surrounding medium, then interference will occur. The resulting small changes in amplitude (brightness) must be exaggerated before the eye can

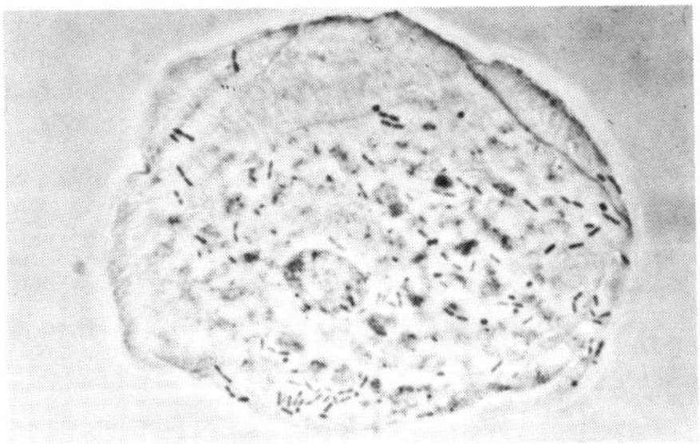

Fig. 109 Cheek epithelial cells under phase contrast. Notice the pronounced peripheral halo. Compare with figs. 32 and 105 (×550).

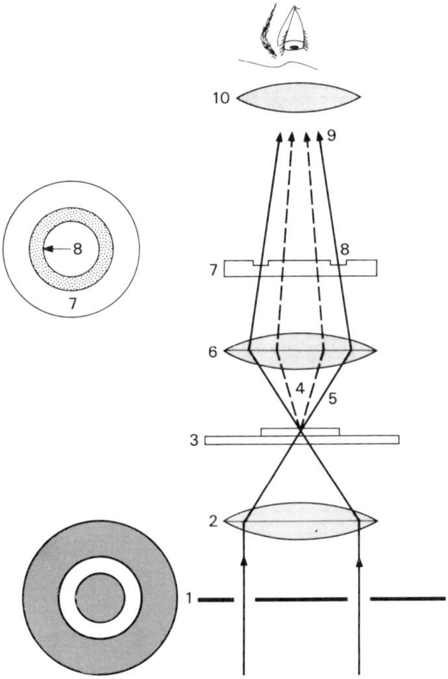

Fig. 110 Phase contrast system as fitted to a light microscope. The lower annular diaphragm (1) produces a cone of light from the condenser (2) which illuminates the specimen (3). Undeviated light (5) (not refracted by the specimen) alone passes through the metal coating of the annular groove (8) of the phase plate (7), which reduces the intensity and advances (quickens) the light waves by $\frac{1}{4}\lambda$ relative to those refracted by the specimen (4), which are slowed approximately $\frac{1}{4}\lambda$ by the specimen. These two sets of waves (9) interfere destructively to produce a high contrast image of the specimen after being integrated by the eyepiece (10). The objective is (6).

perceive them, and this is achieved by a microscope set up for *phase contrast*. If, for example, the specimen slows down the light passing through it by one quarter of a wavelength, then destructive interference will occur if the light is further slowed down optically by a quarter of a wavelength, or if the light passing through the mounting medium is speeded up by the same amount, so that the two waves are half a wavelength out of phase. The specimen will appear dark against a light background (fig. 109).

To produce such changes between light waves, the phase contrast microscope uses a special condenser and special objectives. These can be bought to fit most microscopes but are relatively expensive. The phase contrast condenser has a ring shaped diaphragm (the *annular diaphragm*), which produces a hollow cone of light, from light passing through it. This illuminates the specimen and surrounding medium. The light which passes through the specimen will be *deviated* relative to light passing *directly* through the surrounding medium. The direct light is brought to a focus in the rear focal plane of the objective lens, where a *phase plate* is located. This has a ring-shaped groove on its surface, such that the focused image of the diaphragm annulus and the phase plate groove coincide. The groove, which is coated with a thin layer of metal or metal salt, advances the direct light rays by one quarter of a wavelength and reduces their intensity. The deviated light rays from the specimen which have been retarded compared to light from the illumination source, pass unaffected through the whole of the phase plate. The retarded light waves from the specimen and the advanced light waves from the surrounding medium meet at the primary image plane of the objective lens (fig. 110). Here, some of the light waves will interfere with one another and will be completely extinguished or much reduced in brightness. The eyepiece thus integrates the two sets of light waves to form a highly contrasted image of the specimen. Such an image has a pronounced halo around it because of diffraction effects. The contrast produced by this method depends on there being a difference between the refractive indices of the specimen and the surrounding medium. If these are of similar refractive index, then little contrast will result. For instance, the refractive index of Canada balsam is very similar to that of unstained cells, whereas that of glycerine is not. A glycerine mount would therefore produce better contrast in the specimen. Most phase contrast equipment is designed to work with a green filter, so that light of this wavelength passes through the optical system.

Differential Interference Contrast

Differential interference contrast microscopy eliminates the haloes produced around the images of specimens examined by phase contrast, and

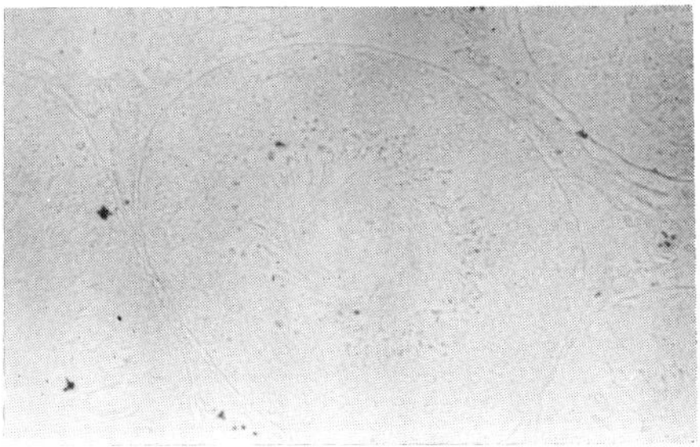

Fig. 111 Unstained TS of mammalian testis in conventional bright field illumination. Very little of the structure of the tubule shown in section can actually be discerned (×200).

Fig. 112 The same section as shown in fig. 111, but observed under differential interference contrast, giving a three-dimensional appearance. Sperm tails within the testis tubule are now more prominent. Compare with oblique illumination (fig. 105) and Hoffman modulation contrast (fig. 114) (×200).

therefore shows the fine detail in the specimen more clearly (figs. 111 and 112). It also introduces colour effects and gives rise to an image with a three dimensional appearance.

Light from the microscope lamp is transmitted through a polariser, which transmits the various wavelengths (colours) vibrating in one plane only (plane polarised light). This plane polarised beam passes through a prism (a birefringent wedge), which separates it into two distinct beams, and rotates the planes of polarisation of these, so that they are at right angles to each other. One of these beams (*the object beam*) travels through some particular feature of the specimen and is slowed down or speeded up. The second beam (*the reference beam*) passes relatively unimpeded through a clear area of the slide. The two polarised beams are recombined by another birefringent wedge located above the objective lens, but they still cannot interfere with one another until they are passed through a second polariser set at 90° to the first and at 45° to each of the incoming beams. This second polariser (or analyser) rotates each polarised beam so that they can interfere with one another (fig. 113).

The second birefringent wedge is responsible for the background colour, which can be varied by moving the wedge in the optical path. The specimen, however, will appear a different colour from the background because of its effect on the object beam. If colour in the object beam is 180° (half a wavelength) out of phase with the same colour in the reference beam, then that colour will be extinguished. By this means some colours will be extinguished while others will be enhanced.

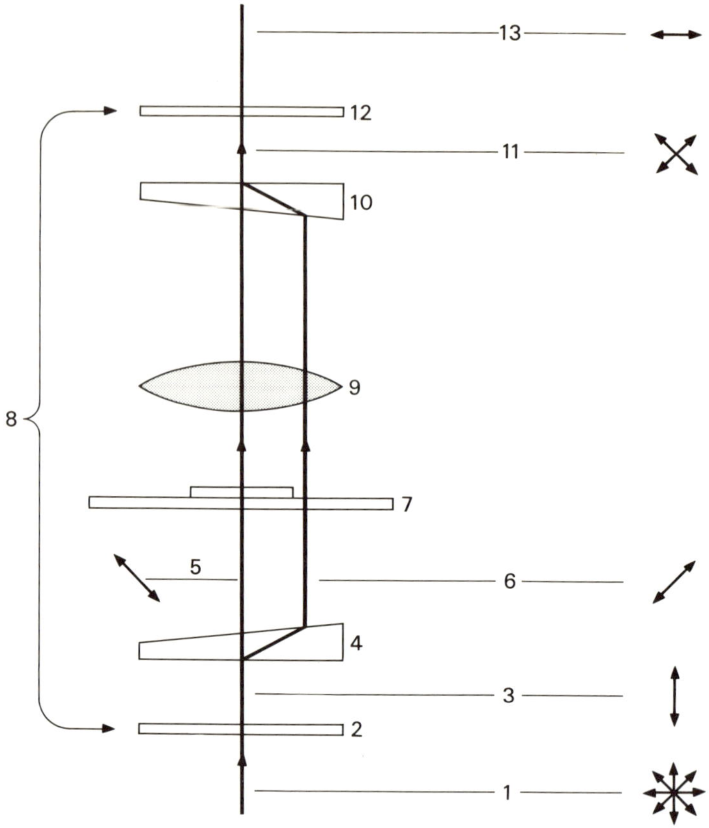

Fig. 113 Features of the differential interference contrast microscope. Light (1) from the lamp vibrates in all directions. The lower polariser (2) produces plane polarised light (3) which is split into two spatially separated beams by the lower birefringent wedge (4). Each of the beams produced is polarised at right angles to the other and at 45° to the original plane polarised beam. One of these beams, the object beam (5), passes through the specimen (7) and is advanced or retarded. The second beam, the reference beam (6), passes through the mounting medium. The two beams pass through the objective (9) and are recombined into a single beam by the upper birefringent wedge (10), which is also responsible for background colour. However the two wave trains (11) cannot interfere with one another until they are rotated into the same plane by the upper analyser (12). Interference occurs to produce the contrast seen in the specimen (13). The polariser (2) and the analyser (12) are crossed (8).

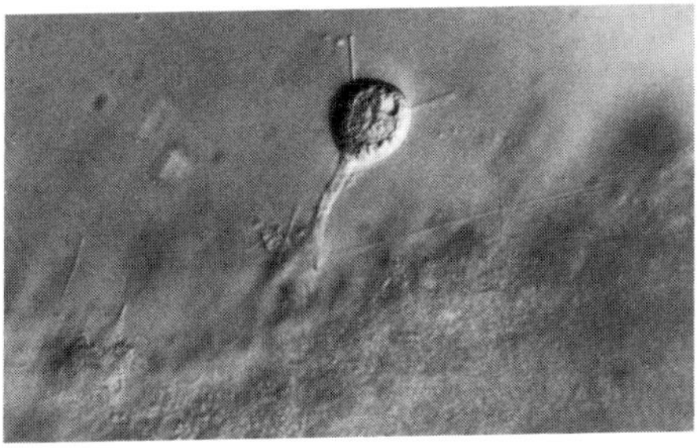

Fig. 114 The suctorian ciliate *Discophrya* under Hoffman modulation contrast, again giving three-dimensional appearance. Compare with figs. 106 and 112 (×120).

Hoffman Modulation Contrast

This contrasting system exaggerates the small differences in refractive indices (phase gradients) between cell components so that they are easily visible through the microscope. Like the oblique illumination and interference contrast methods, the modulation contrast system produces vivid, high resolution, three dimensional images without edge haloes (fig. 114). A major advantage of the Hoffman modulation contrast system is its ability to provide views of any plane within the unstained specimen without obscuring effects from structures above and below that plane. The modulation contrast system converts phase gradients into intensity variations – one phase gradient is rendered dark while the opposite gradient is rendered light. All regions of uniform refractive index appear as grey. The light intensities produced in the image are altered above and below the average grey value, hence the name *modulation contrast*.

These changes are produced by the addition to the microscope of an offset slit aperture, placed beneath the condenser, and a modulator, located in the objective lens barrel, consisting of a dark, a grey and a transparent region. The slit aperture must be aligned with the grey region of the modulator. Light transmitted through the slit aperture illuminates the specimen, and non-refracted light (direct or non-deviated light) is reduced in intensity by passing through the grey portion of the modulator, which reduces the intensity to about 15% of its original value. The light refracted by the phase gradients in the specimen is transmitted either through the transparent portion of the modulator (100% transmission) or through the dark region where the intensity is reduced to about 1% (fig. 115). The eyepiece integrates the light transmitted by the modulator to produce spectacular images.

Autoradiography

Autoradiography is a technique that can be used to determine the location of

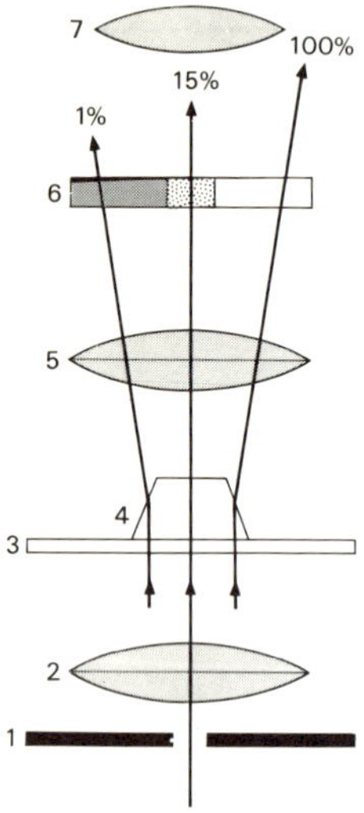

Fig. 115. Features of the Hoffman modulation contrast system. Light passing through a slit (1) beneath the condenser (2) illuminates the specimen (3). Undeviated light passes through the grey region of the modulator (6) where its intensity is reduced (15%). Light deviated by optical gradients (4) in the specimen passes either through the completely transparent region of the modulator (100%), or through the dark region (1%) where its intensity is dramatically reduced. (Percentages show the light transmission.) The eyepiece (7) integrates the rays of light to produce an image of the specimen with increased contrast. (5) Objective.

particular structures or molecules within cells. The living cells are initially fed a component of that structure or molecule in which some of the atoms have been replaced by radioactive isotopes. The cells are then fixed, dehydrated, embedded and sectioned. In a darkroom, a liquid photographic emulsion is coated over the section of cells. These emulsion coated preparations are left in the dark, sometimes for several days or several months, before the photographic emulsion is developed. When the sections are examined under the microscope, the sites, at which the radioactive component has been incorporated into the structure of interest, are shown by silver grains overlying the section (fig. 116). Photographic emulsions are made up of silver halide salt particles embedded in a suitable dispersing medium. They are not only sensitive to light but also to radiation from the decay of radioactive atoms. For example this technique has been used to show the location and sites of production of RNA and DNA in cells.

Fluorescence Staining

Certain dyes emit light when illuminated by short wavelength radiation. Acridine orange is such a fluorescent dye. It can be used in fixed preparations, to detect both DNA and RNA in nuclei and chloroplasts, and to differentiate these structures from the surrounding cytoplasm. In dilute form it can be used to stain living cells; the nuclei stain greenish-yellow and the cytoplasm stains red.

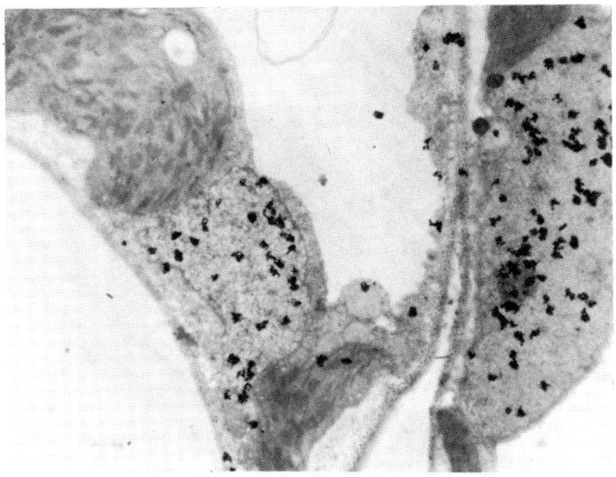

Fig. 116 Electron autoradiograph of a plant section showing silver grains overlying the sites at which radioactive uridine has become incorporated into nucleic acids. This technique is feasible with both light and electron microscopy (×2000).

Fluorescent Antibody Labelling

Fluorescent dyes can be linked to antibodies, and the combination used to locate corresponding antigens in or on cells.

When a foreign protein, for example a virus (*the antigen*), is introduced into the body, a defensive response takes place resulting in the production of a neutralising protein, *the antibody*, which specifically attaches to the antigen to form a harmless complex. This is an important part of the body's defence mechanism against disease. The specific attraction of antibodies to antigens can be used, for example, to diagnose a particular virus infection or to locate various molecules in tissues if the antibody is labelled with a fluorescent dye. This can be achieved by a 'direct' or an

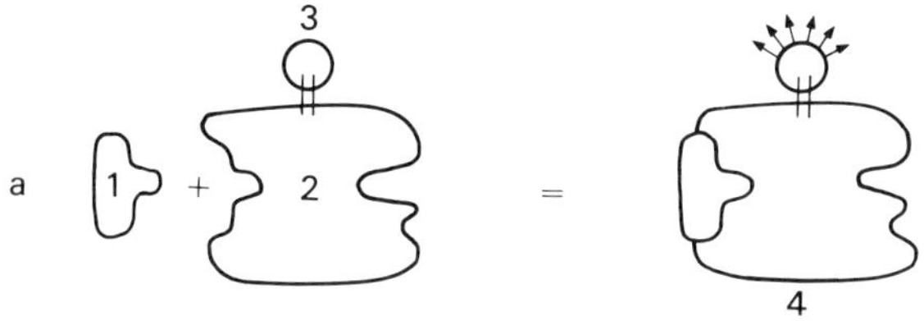

Fig. 117 (a) Direct immuno-fluorescent antibody labelling. (1) Viral antigen. (2) Anti-viral antibody. (3) Fluorescent dye. (4) Fluorescent antigen-antibody complex.

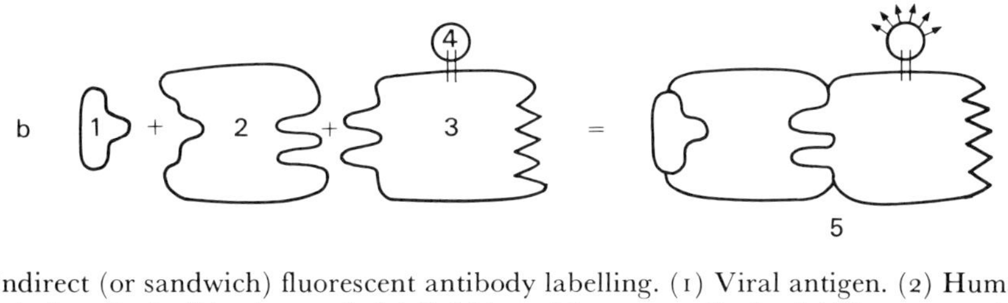

(b) Indirect (or sandwich) fluorescent antibody labelling. (1) Viral antigen. (2) Human anti-viral antibody (blood serum). (3) Rabbit anti-human antibody. (4) Fluorescent dye. (5) Fluorescent antigen-antibody complex.

'indirect' method. In the direct method, the antibody is produced and a fluorescent dye attached. The *fluorescent antibody* is then incubated with cells containing the appropriate antigen, after which, excess antibody is washed away. When this preparation is examined under a fluorescence microscope, the sites of fluorescence will indicate the location of the antigen being studied.

The *indirect fluorescent antibody technique* adds an extra stage to the above procedure (fig. 117), by making use of the observation that human antibodies are treated as antigens by the immunological system of a rabbit (for example), the rabbit would thus produce a specific antibody against the human antibodies (rabbit anti-human antibody). As before, the (rabbit-produced) antibody can be labelled with fluorescent dye and then used to detect the presence of the original type of human antibody; for example, it can be used to discover if a patient has recently had a virus infection such as influenza. In this case, laboratory strains of cell-associated influenza virus would be cultured and then incubated with a blood serum sample from the patient. If influenza antibodies are present in the serum (if the patient has recently been infected) then antibody-antigen complexes will be produced within the virus-infected cells. Excess patient serum can then be washed off and the rabbit anti-human fluorescent antibody added. This will attach to the patient antibody, forming a 'sandwich' of influenza virus, patient antibody and rabbit anti-human fluorescent antibody. After a suitable length of time, excess rabbit antibody can be washed off and the preparation examined in the fluorescence microscope. The presence of fluorescence in the cells indicates that the patient's serum contained antibodies against influenza, and thus the cause of the illness will have been diagnosed (plate 6).

The fluorescence microscope is essentially a conventional transmitted light microscope fitted with a short wavelength radiation illumination source and a dark field condenser, so that the fluorescence from the specimen stands out against the dark background. Oil or glycerine is used between the condenser and the slide. Because ultraviolet radiation is harmful to the eyes, barrier filters are fitted within the microscope.

Recently there has been a trend towards the use of *incident light* fluorescence techniques. In these, short wavelength radiation is directed on to the specimen from above, the objective serving as its own condenser to transmit the radiation to the specimen. The longer wavelength light from the fluorescent dye re-enters the objective and is seen by the eye. Incident light fluorescent microscopes have a *dichroic mirror* in the body tube, this reflects the radiation from the source down into the objective, and thus to the specimen. Light returning from the specimen is not reflected back towards the UV source, but is transmitted through the dichroic mirror,

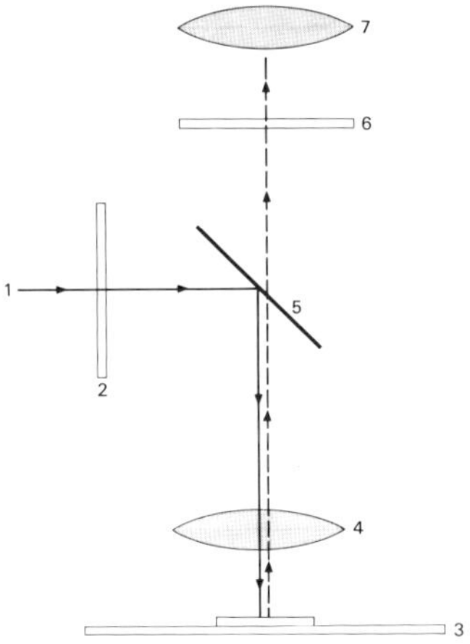

Fig. 118 Principle of incident illumination fluorescence, showing position of dichroic mirror (5), which directs short wavelength radiation (1) from the exciter filter (2) towards the objective (4) from which it is transmitted on to the specimen (3). The returning longer wavelength light is transmitted through the dichroic mirror and barrier filter (6) towards the eyepiece (7).

and into the eyepieces of the microscope (fig. 118). The advantage of this technique is that only the area in view is being illuminated and this increases image brightness. Furthermore, the fluorescent light is not absorbed by the thickness of the specimen, as can happen with transmitted light. This system is convenient to use as it does not require a substage condenser.

Incident Light Microscope

Some materials, such as metals and textiles, are difficult, if not impossible, to make thin enough for light to pass through them. Examination of such materials is usually carried out by reflecting light from their surface into the microscope objective. The surface of solid or opaque specimens, such as coins, can be examined at low magnification in an ordinary compound microscope, if the specimen is brightly illuminated from above. An anglepoise lamp or a bright light source fitted with a condenser is useful for this. Incident light microscopes which work in this manner are used to examine metal surfaces. These metallurgical microscopes are similar in principle to incident light fluorescence microscopes. Light from the lamp is reflected by a semi-silvered mirror into the objective, which acts as its own condenser, to illuminate the specimen. Light reflected back from the specimen forms an image in the microscope body tube, in the usual way, before being further magnified by the eyepiece lens. The light returning through the objective, from the specimen surface, is transmitted through the semi-silvered mirror to form the primary image. The incident light technique has an advantage in that the total aperture of the objective is

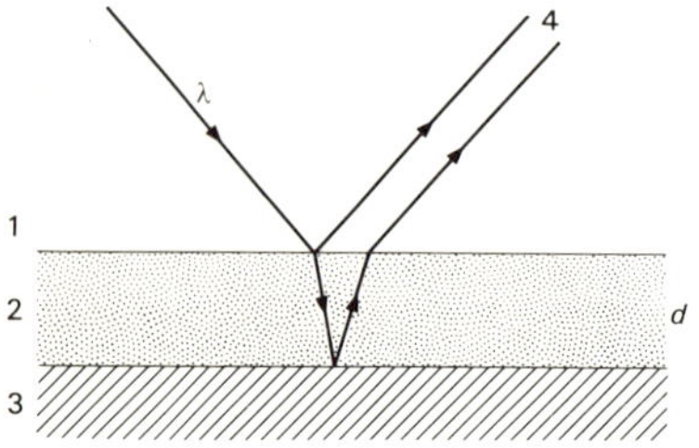

Fig. 119 Mode of action of lens coatings in eliminating reflections which cause loss of image quality in complex optical systems. (1) Air. (2) Lens coating. (3) Glass lens. (4) Destructive interference when reflected rays are a half wavelength out of phase. *d* is the thickness of lens coating.

used, as it acts as its own light condenser. Metallurgists usually examine highly polished metal surfaces, which are etched to reveal the boundaries of the metal crystals or additives. Dark field techniques have been developed for this type of incident light microscopy. A patch stop located in the path of the illuminating beam produces a hollow tube of light. This is reflected by a mirror which has a transparent central area matching the patch stop. Light is transmitted down the outside of special objectives and is reflected on to the specimen as a hollow cone of light. Light reflected from the specimen surface enters the objective, and is transmitted through the transparent centre of the mirror located in the body tube, to form the primary image. The eyepiece further magnifies this to reveal, to the observer, a dark field image of the specimen surface.

Modern Microscope Trends

Some modern research microscopes incorporate many optical developments which may eventually be incorporated into student microscopes (fig. 120). The use of bright quartz halogen bulbs as illumination sources is now fairly common. These are especially useful in photomicrography and in image contrasting techniques, such as phase contrast and interference contrast.

Glass lenses used in the construction of objectives and eyepieces now feature surface coatings (sometimes called *blooming*) of a fluoride salt, such as magnesium fluoride, to increase the amount of light transmitted, and thus increase image quality.

Whenever light hits a lens a small amount is reflected. If there are many lenses in an optical component, as in for example, an apochromatic objective, then this reflected light, on uncoated lenses, provides a background of unfocused light which reduces image clarity. The surface coatings used on the individual lenses of an objective are of a thickness sufficient to cause destructive interference of the reflected light. Some of the light is reflected from the air-fluoride surface, the remainder penetrates the coating and is partially reflected from the fluoride-glass surface.

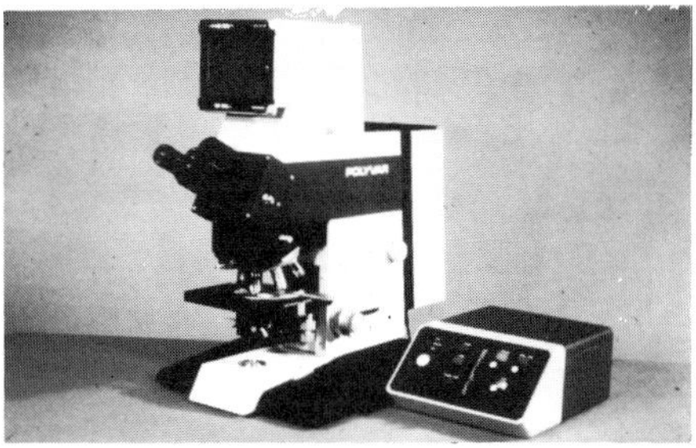

Fig. 120 The Reichert Polyvar research microscope with infinity corrected objectives.

Destructive interference occurs between the two reflected beams because of their phase difference of 180°. Thus, no light is reflected from the lens (fig. 119).

As related in Chapter 1, some of the aberrations inherent in apochromatic objectives are corrected by compensating eyepieces. One manufacturer has recently introduced a new design of objective, where the aberrations are corrected within the objective and these objectives, therefore, do not require compensating eyepieces.

Another development in microscope design is the introduction, by some manufacturers, of microscopes fitted with *infinity corrected* objectives (fig. 120). Here the light leaving the objective is parallel and would thus, theoretically, form an image at infinity, unlike most microscopes where light converges from the objective to form an image in the body tube. Because the light is parallel, many supplementary optical techniques can be performed, by placing a suitable module in the lightpath, without needing a special microscope or special objective lenses. Thus bright field, dark field, phase contrast and interference contrast methods can all be used with such microscopes, simply by changing the module. The tube length of these microscopes is not fixed and may be up to 300 mm long. The image is eventually reformed by an additional lens system, before being magnified in the usual way by the eyepiece lenses. The objectives fitted to these microscopes should not be used on fixed tube-length microscopes.

Clearly, several centuries of improvement have not yet exhausted the potential of the compound light microscope for further development.

9
The Transmission Electron Microscope (TEM)

By the middle of the nineteenth century, microscopists had realised that the light microscope was capable of little further improvement in resolution, and this remained true until the early part of the twentieth century. By this time scientists had found that the limit of light microscope resolution was mainly restricted by the wavelength of the light employed (400–700 nm). A study of potential sources of radiation, of a shorter wavelength than light, and their application to microscopy ultimately gave rise to the electron microscope as we know it. Electrons are very small, negatively charged particles of matter which have wave properties, their wavelength being governed by the velocity of the particles (the faster a beam of electrons, the shorter its wavelength). The wavelength of an electron beam can be many thousand times shorter than the wavelength of visible light. Being negatively charged, the electron beam can also be deflected in magnetic fields, a feature which has given rise to electron optics, somewhat analogous to light optics. Electron microscopy consists of a number of distinct methods, some of which will be described below.

The Design of the Transmission Electron Microscope

Electron microscope lenses consist of a coil of wire shrouded in soft iron (fig. 121). If a current is passed through the coil, a central magnetic field is produced, and this is concentrated by the iron shroud. An electron beam passing through this energised coil can be focused by varying the current

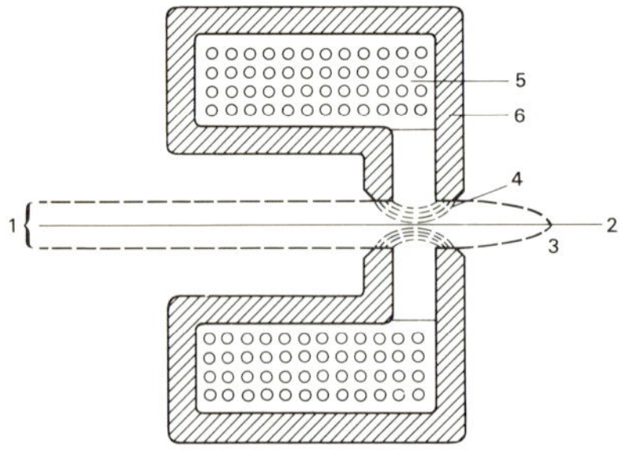

Fig. 121 Section of a magnetic lens used to focus a beam of electrons. A current passed through the coil windings (5) of the lens produces an axial magnetic field (4) which is concentrated by the soft iron shrouding (6) of the coil. (1) Electron beam. (2) Lens axis. (3) Focal plane. (4) Magnetic field. (5) Coil windings. (6) Soft iron.

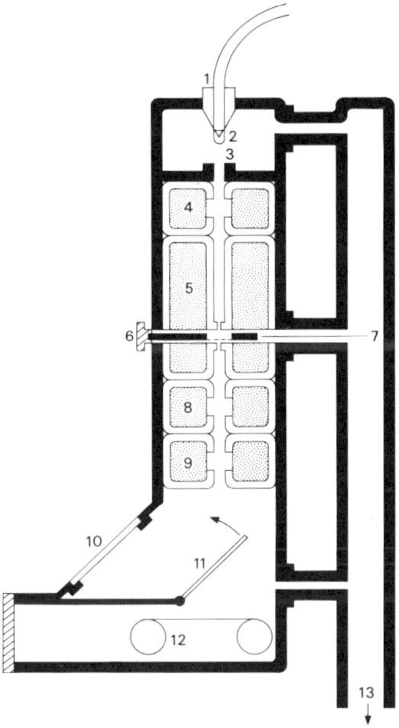

Fig. 122 Section through an electron optical column of a transmission electron microscope showing the major components. (1) High-voltage cable. (2) Cathode containing filament. (3) Anode. *(1), (2) and (3) make up the electron gun.* (4) Condenser. (5) Objective. (6) Specimen door. (7) Specimen rod. (8) Projector I. (9) Projector II. (10) Viewing window. (11) Fluorescent screen. (12) Camera. (13) To vacuum pumps.

passing through the coil. All modern electron microscopes have several of these lenses, which are arranged in a similar way to the lenses of the compound light microscope with a few important differences (fig. 122). The electron path in the electron microscope is evacuated of air, so that electrons are not slowed down by collisions with air molecules. For this reason, the air pressure in the TEM column (the vacuum pressure) is about one ten millionth of atmospheric pressure. It is impractical to move an objective lens up and down, to bring an object into focus, and still maintain a vacuum seal, therefore, the positions of the lenses in an electron microscope are fixed, and the strength of the magnetic field is varied to bring the beam to focus. Electrons cannot be seen, and so a fluorescent screen is used to convert the electron image into a visible light image, which can then be viewed through a lead-glass window at the base of the microscope column. Photographic film is sensitive to electrons so a camera can record the fine detail of this image.

The electron beam is produced by heating a tungsten filament to just below its melting point, this allows electrons to escape from the filament surface. These are accelerated down the microscope column by the application of a very high voltage (30,000–125,000 volts), maintained between a cathode (containing the heated filament) and an anode. The anode has a central hole, through which the accelerated electron beam passes towards the first of the lenses (the condenser), which focuses the beam on to the

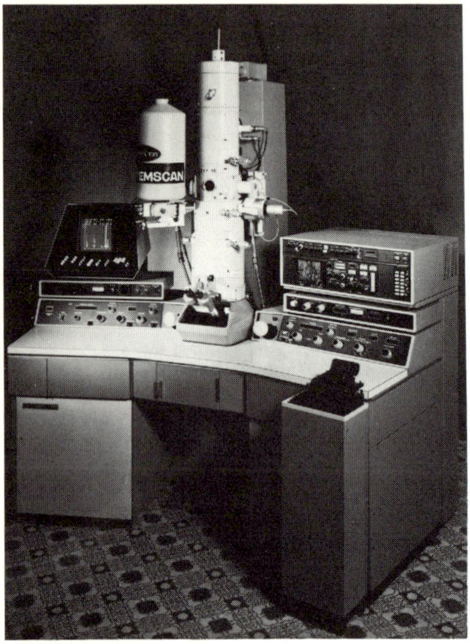

Fig. 123 A transmission electron microscope, the Jeol JEM–100CX, fitted with elemental analysis and scanning attachment (STEM).

specimen contained within the second lens (the objective). The specimen is magnified about a hundred times by this lens, and further magnified by one or more projector lenses (similar to the eyepiece in light microscope optics), before being viewed on the fluorescent screen. Thus the electron microscope produces a magnified image in a similar way to the compound light microscope (fig. 123). The resolving power of a modern electron microscope is of the order of 0·25–1 nm, which is about one thousand times better than the best light microscope.

Like glass lenses, electron microscope lenses have various defects. Chromatic aberration manifests itself in the TEM, when the electrons emerging from the specimen have great variations in their velocities compared to the illuminating beam in which the electrons all have the same velocity. In a TEM the objective lens brings the electrons which pass through the specimen to a focus, but if the emerging beam has a wide variation in speed due to the slowing effect of the specimen (section thickness), then the objective lens cannot bring all the electrons to focus in the same focal plane.

Spherical aberration manifests itself in electron microscope lenses by bringing peripheral electrons, which are influenced by a stronger magnetic field, to a different focus from the central electrons which are influenced by the weaker central magnetic field of the lens.

Specimen Preparation for the TEM

Both biological and non-biological specimens can be examined in the TEM. Sections of such specimens have to be made extremely thin to allow the electron beam to pass through them. This meant that new methods of specimen preparation had to be developed as the light microscope methods were not suitable. The wax embedded specimen used in histology laboratories is of no use in electron microscopy, because wax cannot support the extremely thin sections required. Thus, after fixation (usually glutaraldehyde followed by osmium tetroxide) and dehydration, the specimen is infiltrated by liquid plastic monomers, which are polymerised (in the presence of a hardener and a catalyst, by ultraviolet radiation or heat) to produce an extremely hard plastic block containing the specimen. Sections are cut from this block using an *ultramicrotome* which is capable of producing the ultrathin sections needed (approximately 60 nm thick). Ultramicrotomes are complex, expensive instruments which require a great deal of operator skill (fig. 124). Steel knives with their coarse grain structure are not suitable for cutting ultrathin sections, so glass or diamond knives are used. Glass knives are made by accurately breaking glass and using the freshly broken edge to cut the specimen block. The sections produced by such knives float on water, in a water bath placed around the knife, and their thickness is estimated by interference colours (fig. 125). Gold, silver or grey coloured sections are picked up, on to the surface of 3 mm diameter copper grids which support the specimen (fig. 126). The section of the specimen on the grid is stained using heavy metal salts (uranyl acetate and lead citrate), which impede the path of some electrons in the beam and hence add contrast to the specimen. Coloured stains are of no use in the electron microscope, as electrons are not coloured like the various wavelengths of visible light. The stained sections on the grid are then loaded into the electron microscope, and an image examined on the fluorescent screen at magnifications from about 1000 times to 500,000 times. Biological specimens prepared for electron microscopy are capable of showing detail a hundred times finer than that visible in the light microscope. The resolution of a TEM used for the examination of biological specimens is restricted by the specimen and not by the instrument. Non-biological specimens can, however, allow the full utilisation of the resolving power of the TEM.

Analytical TEM

The electron microscope described above can show structural details within the specimen, but it does not give information about the chemical elements which make up the structures seen on the screen or photographic

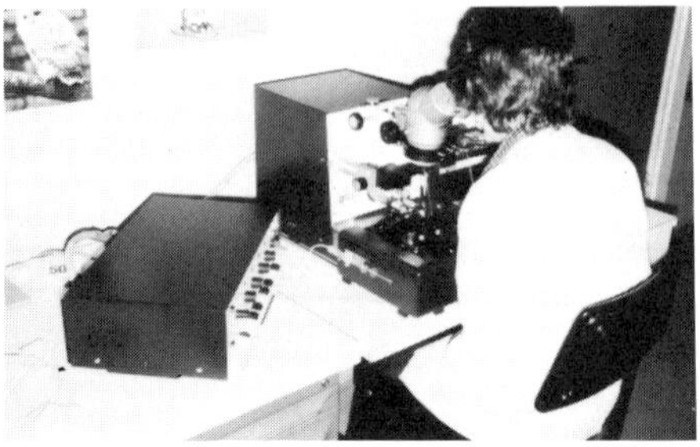

Fig. 124 The LKB IV ultramicrotome in use. (Photograph reproduced with permission of LKB Instruments Ltd.)

Fig. 126 An electron microscope specimen grid. These are normally 3 mm in diameter.

Fig. 125 Close up of an ultramicrotome showing plastic impregnated specimen in block holder (*top*), and glass knife, surrounded by water bath, with ultrathin sections floating on the water surface.

film. A recent innovation in electron microscopy has been the development of *elemental analysis*. This technique allows the detection of many of the elements of which the specimen is composed. The principle behind this analysis results from the interaction of the electron beam and the specimen. The atoms which comprise the specimen have electrons travelling in set orbits (electron shells) around the atomic nucleus. The incident electron beam can displace some of these orbiting electrons with the result that gaps appear in the electron shells. Electrons in outer shells have more energy than electrons nearer the atomic nucleus. Thus outer shell electrons 'jump' to lower orbits if a gap appears. This jump releases a packet of X-radiation. The X-ray emission from a bombarded element is characteristic of that element so, if suitable X-ray detectors are attached to the electron microscope, details of the atoms in the specimen can be obtained. This analytical technique is mainly of use to metallurgists or to biologists examining concentrations of chemicals or crystals found in tissues (for example asbestosis). The technique cannot detect low atomic number atoms (like sodium) or identify organic compounds. A new analytical method potentially capable of detecting these light elements and perhaps organic compounds is under development. This technique, which is called *energy loss analysis*, depends on the examination, by suitable detectors, of the energy lost by electrons in travelling through the specimen.

High Voltage TEM

To achieve acceptable resolution and image illumination, the transmission electron microscope previously described can only be used to examine extremely thin specimens. This is due to the limited penetrating power of the illuminating electron beam. To examine thicker specimens a more energetic beam of electrons is required. Such a beam can be produced by increasing the velocity of the electrons by using a higher electron gun accelerating voltage. The generation of highly stable voltages up to and in excess of one million volts cannot be achieved cheaply, nor can the high voltage generators be kept compact. Thus high voltage microscopes are large and costly (fig. 127). It is therefore not surprising that, throughout the world, few microscopes of this type have been installed. High voltage microscopes are used mainly by metallurgists, by scientists who wish to examine specimens incapable of being made thin enough for a conventional transmission electron microscope, or wish to observe radiation damage in specimens. This last use is important, as materials used in the construction of nuclear reactors need to be tested. Few biological specimens have been examined in the high voltage electron microscope, although such microscopes may usefully be used to examine whole cells. The interpretation of the images produced by such relatively thick specimens as whole cells is problematical, as cell structures overly one another. However, the application of three dimensional imaging may give cell biologists a new insight into the relationships between cell organelles.

Viruses, Bacteria and Cells under the TEM

This section describes some of the results obtained from the transmission electron microscope in elucidating the structure of viruses, bacteria and cells. Cells are seen relatively easily under the light microscope, but sub-cellular structures are not. Bacteria are visible under the highest magnifications of the light microscope, but only the largest of viruses can be seen in light microscopes set up to attain the highest resolution. However, the resolution available in the electron microscope has enabled

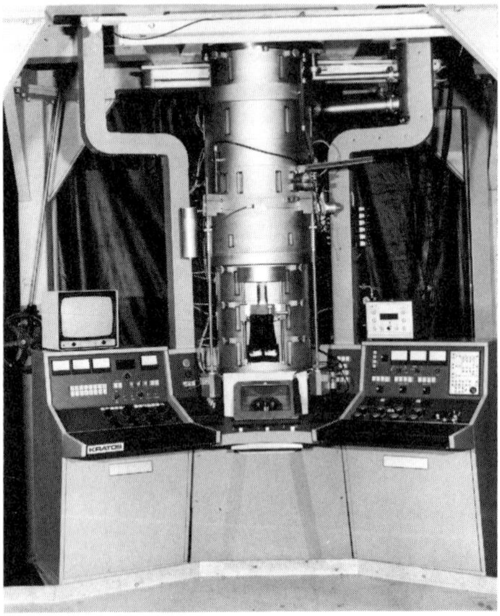

Fig. 127 A high voltage transmission electron microscope, the Kratos EM7.

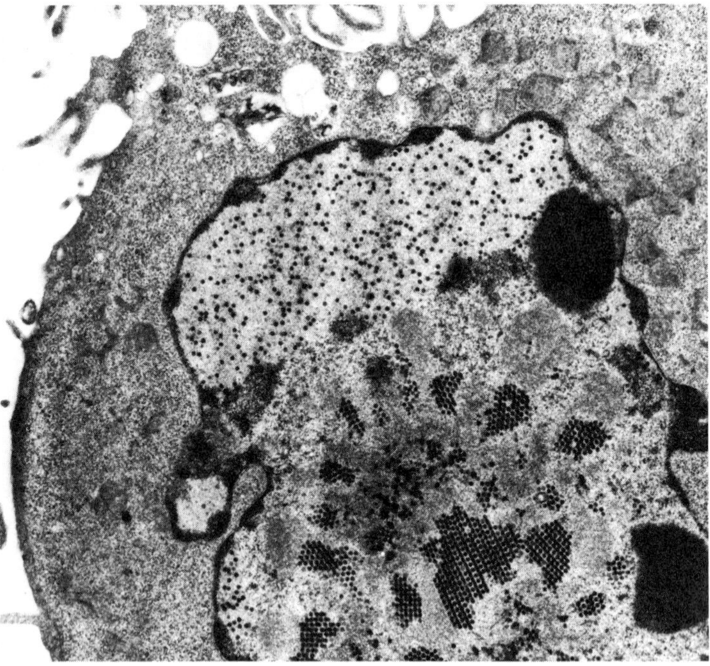

Fig. 128 Section of an adenovirus-infected cell, showing numerous viruses in the nucleus (×8700).

scientific workers to reveal a remarkably beautiful and complex world of sub-cellular biology.

THE STRUCTURE OF VIRUSES

Viruses were among the first biological subjects to be examined in the electron microscope. They are obligate intracellular parasites, requiring living cells in which to replicate (fig. 128). This often causes illness, when cells or organisms are infected. Viruses are extremely small, and are generally composed of a symmetrical protein shell containing a nucleic acid (either DNA or RNA). They show a great variety of structure and size. Viruses are classified or grouped by comparison of their size, symmetry and nucleic acid composition. The largest viruses (poxviruses) are just visible under the best light microscopes but their shape cannot be made out. However, in the electron microscope they are resolved easily, indeed certain virus diseases are diagnosed by electron microscopy. Various groups of viruses are known, these vary in diameter from 20 nm to 250 nm (figs. 129 and 130). Viruses are prepared for examination with the electron microscope, by allowing a virus suspension to dry on to a specimen grid coated with a suitable support film transparent to the electron beam. A heavy metal salt solution (uranyl acetate, potassium phosphotungstate, etc) is used to stain the virus particles. The stain dries around the particle and inhibits transmission of the electron beam immediately around the particle, and thus the particle is easily seen against the background. This is called *negative contrast staining*.

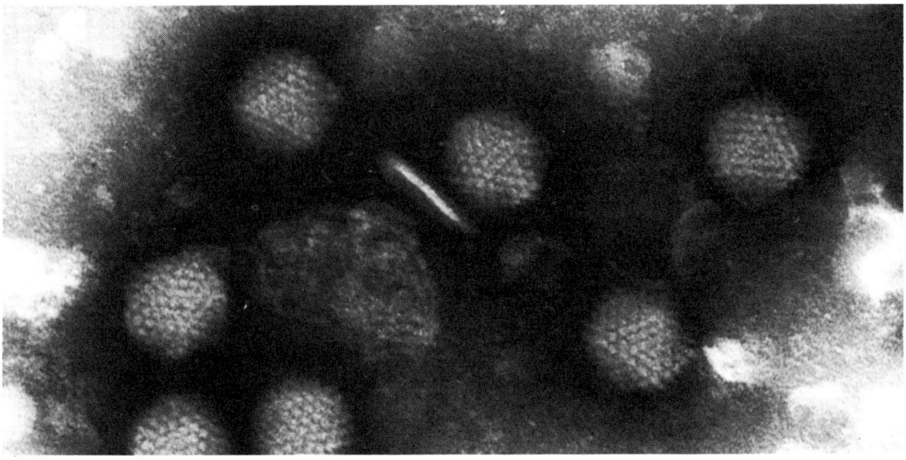

Fig. 129 Negatively stained adenoviruses, showing hexagonal outline and granular appearance due to the units *capsomeres* which make up the protein shell (×200,000).

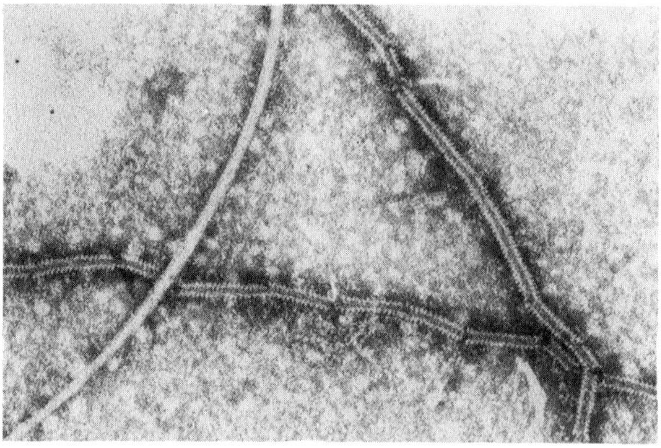

Fig. 130 Negatively stained preparation of a helical-symmetry virus consisting of a long protein tube enclosing the nucleic acid. A featureless bacterial flagellum can be seen overlying the viral structures (×110,000).

BACTERIAL STRUCTURE

Unlike viruses, bacteria are capable of an independent existence, providing they have suitable conditions and nutrients in which to grow. They, like viruses, can cause disease, but they also cause spoilage of food and can prove a nuisance by contaminating fuels. However, the vast majority of bacteria are not harmful to man, and are useful in such diverse areas as water purification, cheese production, waste breakdown, enzyme production for biological washing powders, or synthesis of useful organic molecules. Compared to the complex internal structure of the cells of which our bodies are composed, bacteria seem simple. They average about 1 μm in diameter, and have a cell wall surrounding a cell membrane which encloses the granular bacterial cytoplasm and filamentous nucleic acid. They may, however, have structures present on their cell wall which are easily seen by electron microscopy. These include flagella which are not in any way similar to the flagella or cilia found in protozoa, algae or some animal cells (fig. 131). The current theory of bacterial flagella motion is that the filament rotates in the bacterial wall and cytoplasmic membrane, analogous to the rotation of an electric motor. This type of rotatory movement is unique to bacteria. Other surface structures found in some bacteria are pili or fimbriae, which are very thin (sometimes rigid) tubular structures (fig. 132). Their true function is uncertain but they may be involved in attachment to unrelated cells, this being an important feature of cell infection. They also transfer genetic information between one bacterium and another, which is important as antibiotic-resistant bacteria can transfer this resistance, via sex pili, to other bacteria. Some bacteria also have a polysaccharide capsule around them which can protect them and aid their attachment to surfaces (fig. 133). This last aspect is particularly important, as dental plaque is an assemblage of bacteria embedded in capsular material, which is attached to teeth, and thus allows the bacteria to release digestive enzymes which ultimately

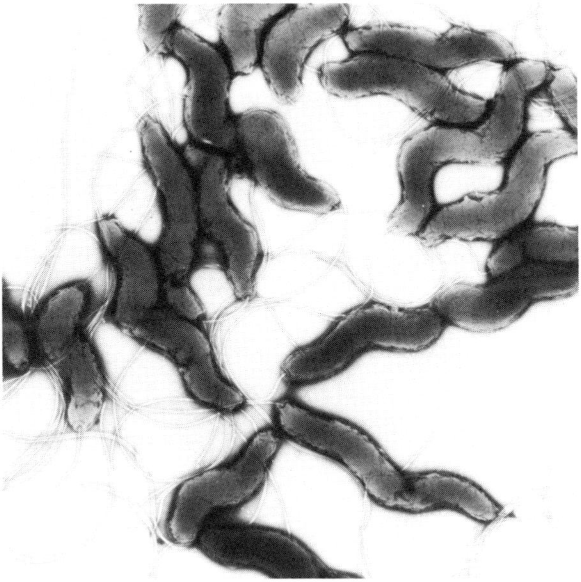

Fig. 131 Negatively stained bacteria *Campylobacter* showing terminal flagella (×6000).

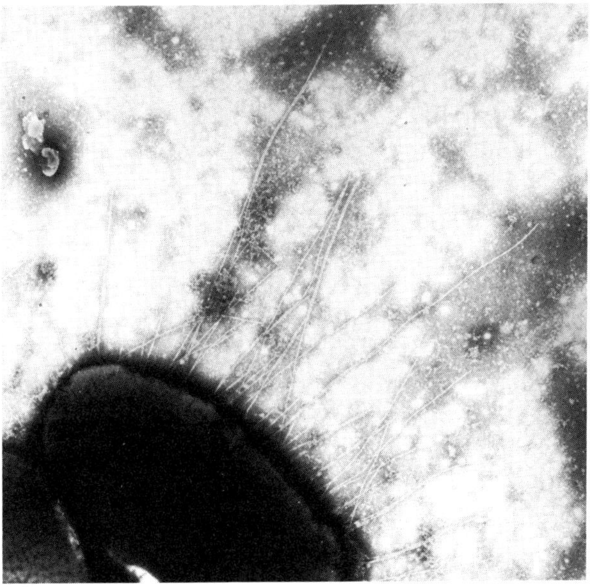

Fig. 132 Negatively stained bacterium *Shigella* showing radiating long thin surface structures (*pili*) (×27,500).

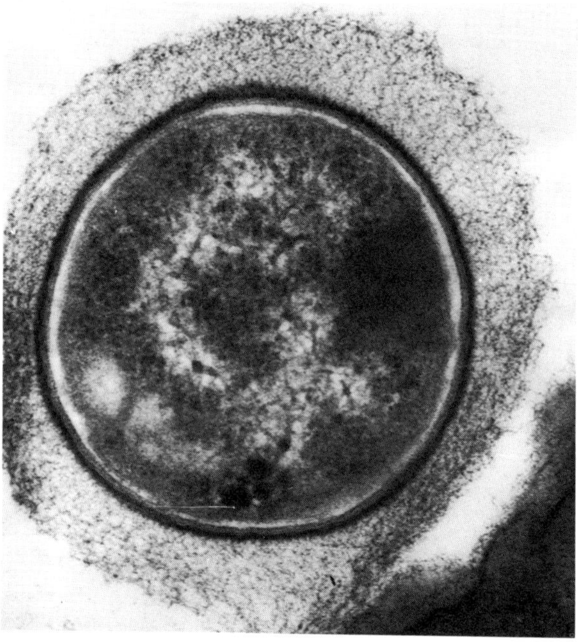

Fig. 133 Thin section through the bacterium *Klebsiella* showing external fibrous capsular material. Also note internal filamentous zones which indicate the position of the bacterial chromosome (×80,000).

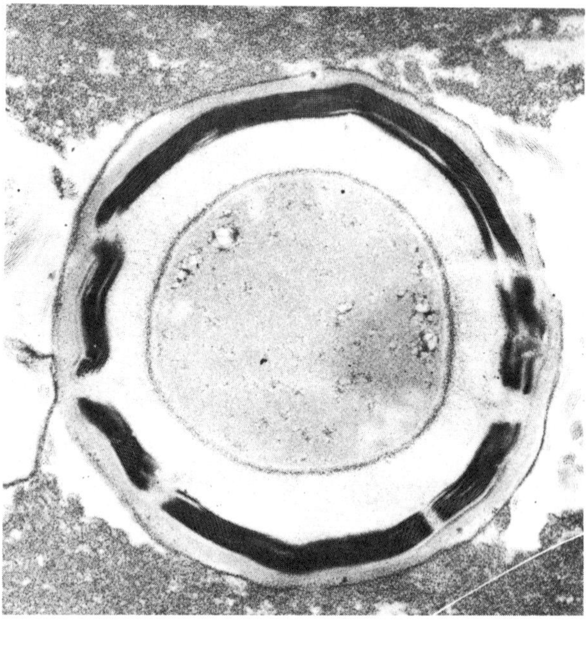

Fig. 134 Section of germinating bacterial spore, obtained from lung material from an Egyptian mummy. Note complex concentric coats surrounding central bacterial cell (×70,000).

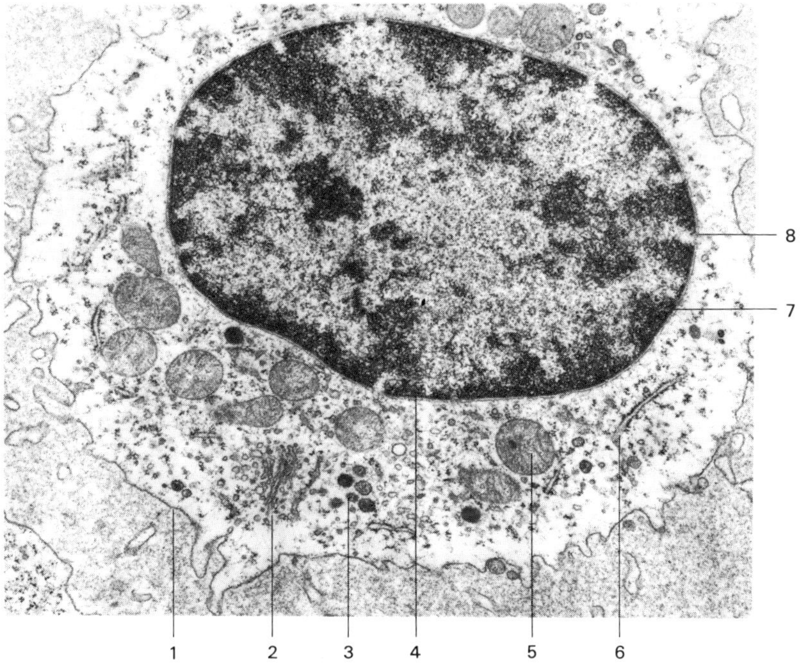

Fig. 135 Thin section of a human white blood cell (*monocyte*) showing typical features of an animal cell (×5000).
(1) Plasma membrane. (2) Golgi apparatus. (3) Lysosome. (4) Chromatin (5) Mitochondrion. (6) Rough endoplasmic reticulum. (7) Nuclear membrane. (8) Nuclear pore.

cause dental decay. Some bacteria produce highly resistant spores of complex structure which can resist boiling (fig. 134). Spores are produced when adverse conditions prevail in the bacterial environment.

ANIMAL AND PLANT CELLS

Animal and plant cells contain many structures which have only been properly revealed by electron microscopy. Unless special stains are used, living cells examined by light microscopy only show the nucleus and the movement of tiny granular cytoplasmic structures. Cells examined in the electron microscope show that the granular cytoplasm seen by light microscopy is composed of many organelles, which differ in structure and function. A generalised view of cell structure will be described here, but it must be understood that cells which make up body organs and plant tissues can differ in detail. Cells are surrounded by an external membrane, the plasma membrane, which separates the internal components from the environment outside (fig. 135).

The most obvious structure in most cells is the nucleus, which contains the genetic store of that cell and of the whole organism. The nucleus is surrounded by an envelope composed of two membranes, often punctured

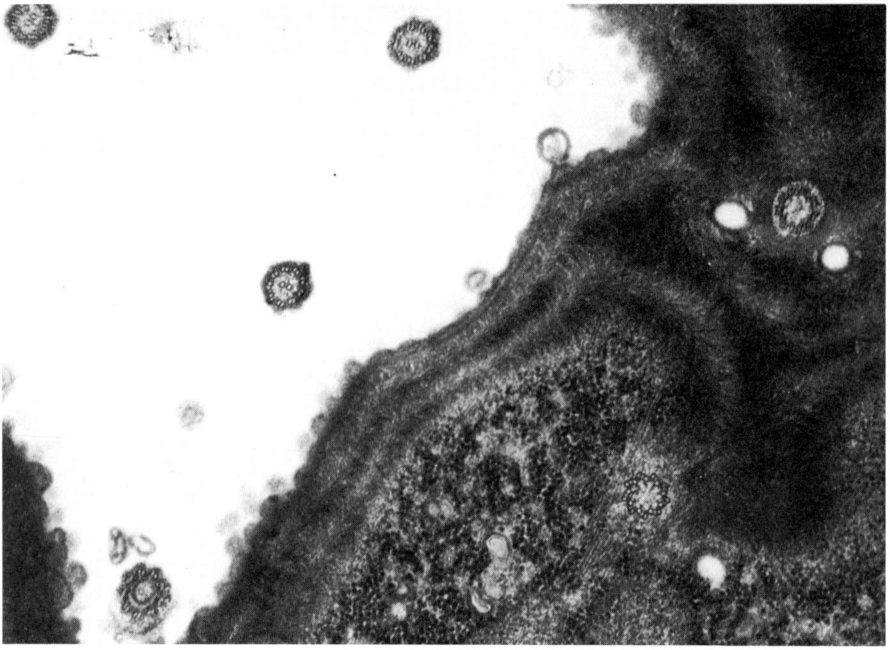

Fig. 136 (a) Section through the surface of the ciliated larva of the suctorian *Discophrya* showing 9+2 arrangement of microtubules in ciliary shaft, and 9 triplets of the basal body. The latter structure is identical with a centriole (×30,000). (b) Longitudinal section of cilia from the feeding organelle of the ciliate *Blepharisma* (×15,000).

by pores. These allow passage of materials to and from the surrounding cytoplasm. The genetic material normally appears scattered throughout the nucleus, except just prior to cell division when it condenses to form individual chromosomes. Often a body called the nucleolus is found in the nucleus and this is largely composed of RNA, a vital component in the translation of the DNA into the messages which govern the productive machinery of the whole cell.

The most obvious cytoplasmic components of cells are the mitochondria. These organelles are usually sausage-shaped, and are composed of a smooth outer membrane which surrounds an internal convoluted membrane. The convolutions of the internal membrane are called *cristae*, and on them are located the respiratory enzymes. Mitochondria ultimately convert the energy of digested food in to a form (ATP) which the cell can utilise. Thus mitochondria have been called the powerhouses of the cell.

Another cell component, which is often striking under the electron microscope, is the endoplasmic reticulum (ER). This usually occurs in two forms, the rough ER and the smooth ER. Rough ER is a system of membrane walled canals. The walls of these canals are studded with ribosomes, which are small granular structures composed mainly of ribosomal RNA (fig. 135). Ribosomes assist in the process of the translation of genetic instructions from the nucleus, in the form of messenger RNA, into the production of proteins, which are stored or separated from other cell components within the walls of the endoplasmic reticulum until required.

Another organelle which is often found in association with the ER is the golgi apparatus. This is composed of a series of stacked, flattened membrane sacs and spherical vesicles (fig. 135). The golgi apparatus is thought to package cellular products for external secretion and to form lysosomes. Lysosomes are packets of hydrolytic enzymes responsible for digesting cell organelles no longer required by the cell, or for digesting cell invaders such as bacteria. On the death of a cell, lysosomes are responsible for digestion of that cell, a process called autolysis.

The cell contains a number of other organelles, which are easily seen in the electron microscope. Centrioles dictate the position and form of the mitotic spindle during cell division. Centrioles have a complex, but symmetrical appearance (fig. 136). They are tubular in form, with walls composed of nine overlapping triplets of smaller tubules (microtubules). Centrioles are unusual in that they replicate themselves by forming a small daughter centriole at right angles to the mature organelle.

Microtubules are commonly found in cells, but are most noticeable during cell division, as the spindle is predominantly composed of these. Their normal function in the non-dividing cell is a cytoskeletal one. In

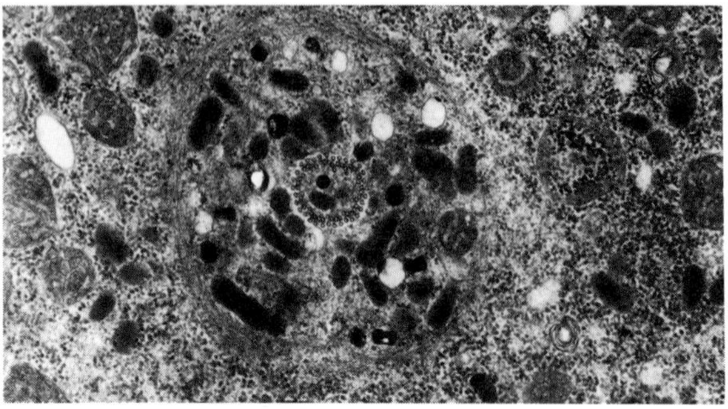

Fig. 137 Section through the body of an adult suctorian ciliate showing transverse section of tentacular microtubules and surrounding microfilaments (×18,000).

this role, they maintain cell shape and are frequently found beneath the cell membrane. Microtubules are also important components in those locomotory organelles capable of moving the whole cell or capable of producing external fluid motion (cilia, flagella and sperm tails). Cilia and flagella have a membrane envelope surrounding a central core made up of nine pairs of microtubules, which form a ring around two single microtubules. The entire microtubular complex of cilia and flagella is called the *9 + 2 structure* (fig. 136). Movement of the cilium is by a mechanism based on the sliding of these microtubules relative to one another.

Fine filaments are also found in cells. These microfilaments are composed of a protein (actin), and are involved in certain types of whole cell movement (amoeboid movement), daughter cell separation following chromosomal segregation and movement of cell organelles within the cytoplasm (fig. 137). Muscle cells are highly specialised cells which contain bundles of microfilaments. The actin interacts with another protein (myosin), which causes these two types of protein to slide relative to one another. This sliding mechanism facilitates the shortening of the muscle cell which produces the muscle movement.

The Freeze-Fracturing Technique

Certain electron microscope techniques, such as *freeze-fracturing*, do not require chemical fixation. Freeze-fracturing requires rapid freezing of the specimen (*cryofixation*) to ensure there are no ice crystal artefacts, that is, structures which are not of biological origin, but are a consequence of the preparative technique. The frozen specimen is then fractured (it can also be sectioned at this stage) and some ice allowed to sublime off as a vapour to reveal certain structures. A carbon replica is then reproduced of the specimen surface. This replica is detached and is metal coated from an angle to produce contrast (*shadowing*). Examination of such a specimen in the electron microscope allows the visualisation of membrane surfaces which have not been modified by the chemical fixatives used in other techniques.

10

Scanning Electron Microscopes (SEM and STEM)

The scanning electron microscope (SEM) is a complementary instrument to the transmission electron microscope described previously (fig. 138). Under the transmission microscope we can study the finest details of cell structure in thin sections. However, with the SEM we can examine the surface details of a specimen.

The electron optical column of the SEM is shorter than that of the TEM and contains an electron source (electron gun), a lens system for focusing the electron beam to a small spot, and deflection coils for sequentially

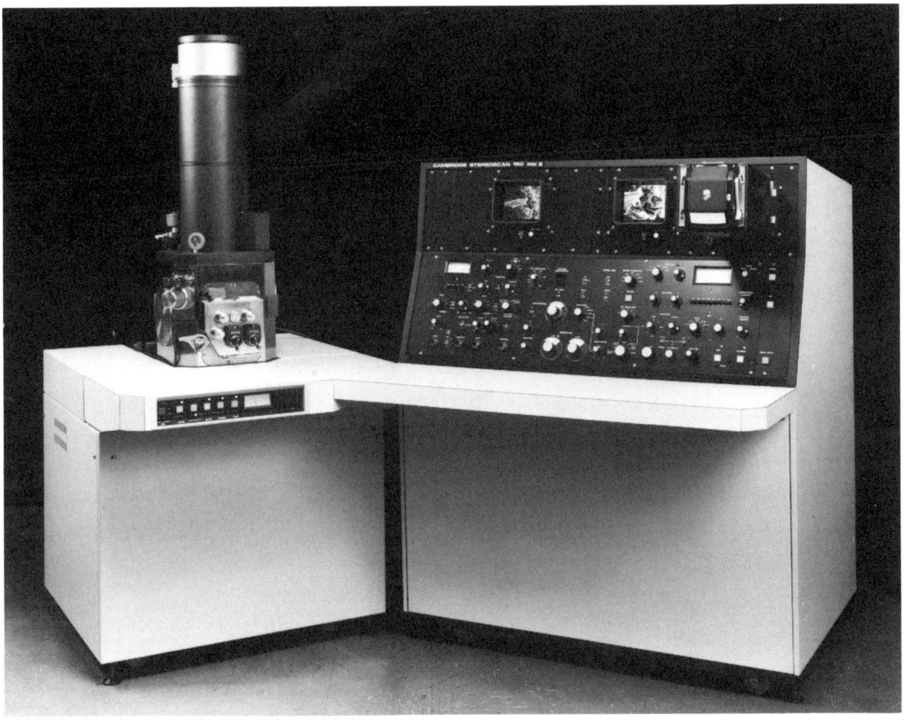

Fig. 138 A scanning electron microscope, the Cambridge Stereoscan 150 Mark 2. Note cathode ray monitors of right-hand module displaying the final image of the specimen.

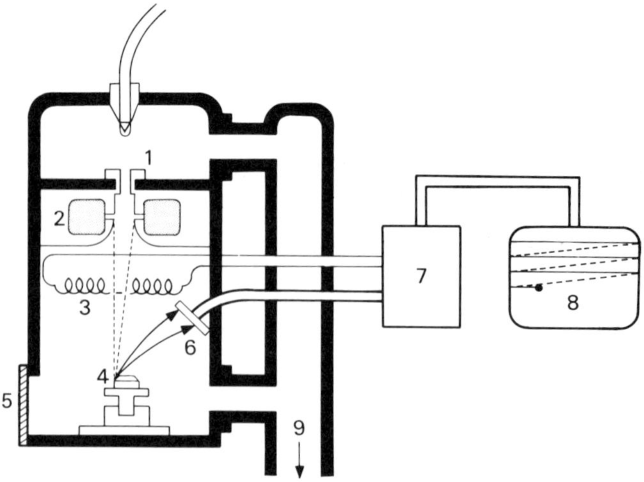

Fig. 139 Diagram through the column of a scanning electron microscope, showing position of the lens (2), scanning coils (3), specimen (4) and reflected-electron detector (6). The electrons emitted by the specimen are processed electronically and an image of the specimen surface is displayed on a cathode ray tube monitor (8). The electron beam scanning the specimen is in exact synchrony with the beam producing the magnified image on the cathode ray tube monitor. (1) Electron gun. (2) Lens. (3) Beam scanning coils. (4) Specimen. (5) Specimen door. (6) Electron detector. (7) Electronics. (8) Cathode ray tube, showing sequential image build up. (9) To vacuum pumps.

moving this fine beam of electrons over the specimen, and also the electron collector. The beam of electrons is made to scan the surface of a specimen, which emits secondary electrons (fig. 139). These secondary electrons, of various velocities, are collected by a detector (*scintillator*), the function of which is to convert the kinetic energy of the electrons into a current. This current modifies the intensity of a beam scanning the screen of a cathode ray oscilloscope in exact synchrony with the beam scanning the specimen. By this means, an image is built up sequentially, each point on the surface of the specimen being scanned corresponding to a point on the screen image. Image contrast and brightness are controlled electronically, like a television image. The magnification of the specimen is the ratio of the area being scanned relative to the size of the screen used to view the image. The larger the area of the specimen scanned the smaller the magnification on the viewing screen, and conversely, the smaller the area scanned the greater the magnification of the specimen. The resolution attained by the SEM is about one-tenth that achieved by the TEM (10 nm SEM, 1 nm TEM).

Specimen Preparation for the SEM

Metallic specimens, or specimens which are resilient enough to withstand

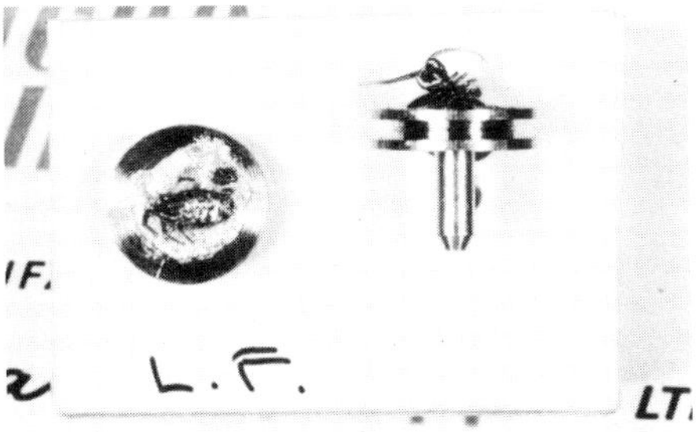

Fig. 140 Two views (top and side) of the specimen holder (*stub*) used for examination of specimens in the scanning electron microscope. The mounted specimens are covered in a conductive metal coating (×1).

collapse under vacuum conditions, can be quickly loaded into the SEM column, the column evacuated and the specimen examined. However, many biological specimens, because of their soft nature, collapse in the vacuum of the microscope. Such specimens have to be made resilient to collapse by methods such as *freeze drying*, where the water content of the specimen is removed without the structure of the specimen collapsing. After freeze drying, biological specimens are usually coated with a thin layer of gold by a vacuum evaporation procedure. This coating removes excess heat and electrical charge from the specimen while it is scanned by the electron beam (fig. 140). SEM images are visually spectacular, as the illustrations show, and provide high resolution images of specimen surfaces, with an excellent depth of focus (figs. 141, 142 and 143).

The Scanning Transmission Electron Microscope (STEM)

In the previous chapter a short description of the high voltage TEM was given, and its application to viewing slices of specimens too thick to be examined in a conventional TEM was described. In recent years several manufacturers have developed the *scanning transmission electron microscope* (STEM), which can be used to examine thick specimens. In addition, STEM instruments can be used to look at conventional ultrathin sections, and surfaces of specimens.

Two types of STEM are available, the first is a conventional TEM with a scanning attachment fitted. Such (relatively inexpensive) instruments are becoming increasingly common in electron microscope laboratories (fig. 123). The second type of STEM is a purpose built microscope without the inherent limitations brought about by converting a conventional TEM. These instruments are much more expensive than the converted transmission microscopes and very few have been installed in laboratories worldwide. The STEM image is built up by scanning a small diameter beam of electrons over the section of the specimen under examination, and using the transmitted electrons to control a cathode ray

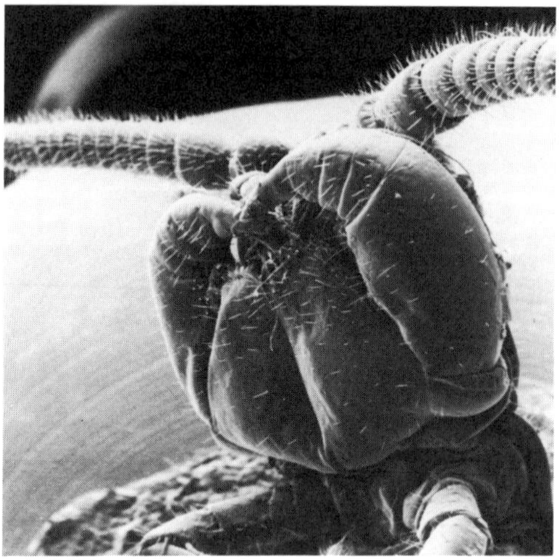

Fig. 141 Scanning electron micrograph of the ventral surface of the head of the centipede *Lithobius forficatus*. Particularly prominent are the poison claws (×20).

Fig. 142 Scanning electron micrograph of the grain weevil *Sitophilus* (×40).

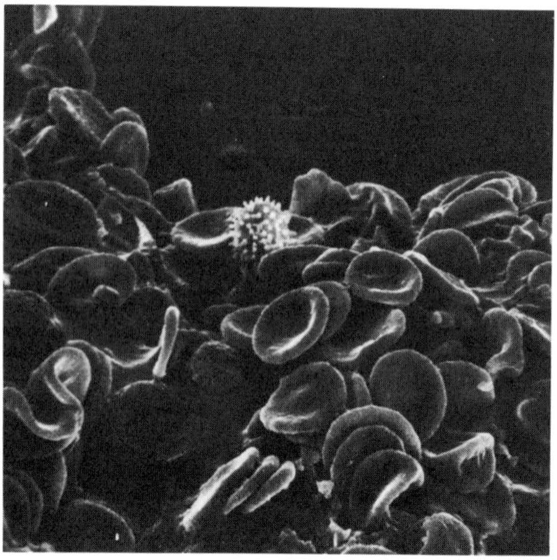

Fig. 143 Scanning electron micrograph of human blood, showing numerous biconcave red cells and a single white cell with surface protrusions (×2000).

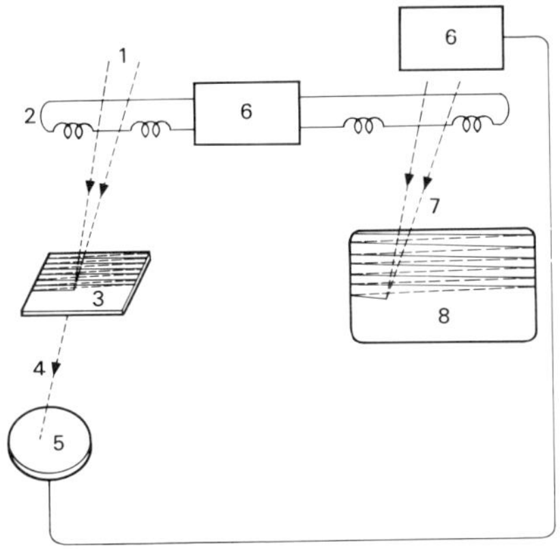

Fig. 144 Diagrammatic representation of the scanning transmission electron microscope (STEM). (1) Electron beam. (2) Electron beam scanning coils. (3) Section of specimen. (4) Transmitted electrons. (5) Electron collector. (6) Electronics. (7) Cathode ray tube electron beam. (8) Image displayed on cathode ray tube.

tube display, as related earlier in SEM image formation (fig. 144). Again, each point on the display corresponds with a point on the specimen. The resolution of the image produced is primarily governed by the diameter of the beam used to scan the specimen and not by the lens defects found in TEMs. Thus, to compete with the resolving power of the transmission microscope the beam must have a diameter of the order of 1 nm or less. The electron beam produced from a conventional tungsten filament cannot achieve the brightness required at such small beam diameters, so a new type of electron gun is used, the *field emission* electron source. These electron guns utilise a fine pointed crystal of tungsten which is placed in an electrostatic field. Electrons are pulled from the surface of the tungsten crystal by this field, which produces an intense electron source up to one hundred thousand times brighter, and a thousand times smaller, than a heated tungsten filament can produce. However, the vacuum environment required by this electron gun has to be one million times better than that employed in conventional transmission and scanning electron microscopes. This means that special vacuum pumps and seals have to be used in the construction of the STEM instrument. Electrons produced by this source are accelerated towards a lens used solely to focus an intensely bright electron beam, 0·2–0·5 nm in diameter, on to the specimen. This electron beam is scanned sequentially across the specimen. There are no objective or projector lenses after the specimen, so the spherical and chromatic aberrations of the imaging lenses of the TEM are eliminated.

STEM instruments can also be used for chemical analysis of the specimen, as described previously. Scanning transmission instruments combine several electron optical techniques into a single microscope and thus are very versatile.

Appendix I: Materials

Further details of most of the chemicals referred to in the text are given here. Stains, fixatives, and so on, are usually obtainable ready-made from the suppliers listed in Appendix II, but the constituents can also be purchased, either from the same suppliers or from pharmacists, so instructions for making these materials are given below.

Many of these chemicals carry a significant safety hazard, either through their toxic or their inflammable nature. This means that suppliers, while perfectly willing to deal with private individuals, are unable to supply those individuals with certain chemicals. Those that can be supplied may still be potentially hazardous, and it is therefore essential to adopt standard safety practice when dealing with all chemicals. Store chemicals away from heat in stoppered bottles, wipe up all spillages immediately, and wash away spillages on the skin with copious amounts of water. Where a technique involves boiling, this should always be done gently, preferably in a fume cupboard, or at least near an open window.

Dilutions

These are usually expressed as a percentage solution in a particular solvent, generally distilled water or alcohol. As an example, a 60% alcohol solution would be made up by mixing 60 cm^3 of absolute alcohol with 40 cm^3 of distilled water, or any other volumes kept in a 6:4 proportion. Some dilutions are expressed as 1 in 10,000. To make up such a solution, dissolve 0·1 g of material in 1 litre of water.

Mounting Media

Choice of mounting medium depends partly on the medium's refractive index (n), which influences the clarity of the mounted object, and partly on the context of mounting (see relevant chapters).

a) 50% **glycerine** ($n = 1{\cdot}40$), made by adding equal parts of glycerine and water.

b) **Glycerine jelly** ($n = 1{\cdot}42$). Soak 10 g of gelatine in 60 cm^3 of distilled water for about 2 hours, then add 70 cm^3 of glycerine and 0·23 g of phenol. Heat the mixture gently until it is smooth, then store in a refrigerator. Apply the mountant in melted form.

c) **Canada balsam** ($n = 1{\cdot}43$ approx). The most widely known permanent mountant, with a refractive index similar to glass. Xylene is a suitable solvent, either for thinning or for removing spillages. Geological mounts are made in natural filtered Canada balsam which is not in solvent.

d) **Euparal** ($n = 1{\cdot}48$). Similar to Canada balsam, but specimens can be mounted directly from 95% alcohol.

e) **Naphrax** ($n = 1{\cdot}77$). A high refractive index resin dissolved in toluene. Once the specimen is mounted, the toluene must be driven off by very gentle heating of the slide. Although toluene is inflammable, the minute amounts involved do not pose a significant safety hazard. Available from Northern Biological Supplies.

f) **Plastic Padding** super epoxy resin ($n = 1{\cdot}56$). A transparent resin supplied by hardware shops as two tubes, the contents of which must be mixed. Similar resins may be equally suitable for rock thin sections provided they tightly bond the section to the slide.

g) **Lakeside** 70 °C thermoplastic mounting cement ($n = 1{\cdot}54$) for rock thin sections.

Stains

Stains are listed here in alphabetical order – for details of staining techniques, see the relevant chapter. Alternative quantities and methods can be found in other books.

a) **Acetic orcein**. Use 1–4 g of orcein (quantity is variable depending upon variations in the dye, and can only be determined by trial and error) in 100 cm^3 of 45% acetic acid in a flask, and gently boil for 30 minutes, in such a manner as to minimise water loss. When cool, store in a refrigerator. Use the stain with dilute hydrochloric acid in the proportion 9:1 (stain:acid).

b) **Aniline sulphate**. Use a saturated solution in water, with a few drops of dilute hydrochloric acid added.

c) **Basic fuchsin**. For Gram's staining method use basic fuchsin as a 0·1% solution in distilled water.

d) **Borax carmine**. Dissolve 4 g of borax and 3 g of carmine in 100 cm^3 of distilled water. Boil for 30 minutes, allow to cool, and add an equal volume of 70% alcohol. Filter the stain before use.

e) **Congo red**. Dissolve 0·1 g of congo red in 100 cm^3 of 20% alcohol. Filter before use.

f) **Crystal violet**. Use a 1% solution of crystal violet in distilled water, and mix with a 5% solution (in distilled water) of sodium bicarbonate. Mix in a ratio of 30 volumes of crystal violet to 8 volumes of sodium bicarbonate.

g) **Gram's iodine**. Add 1 g of iodine and 2 g of potassium iodide to 25 cm^3 of water, stir, and make up to 300 cm^3 with distilled water.

h) **Haematoxylin** and **eosin**. Use eosin Y as a saturated solution made up in 90% alcohol. Haematoxylin solution can be made up by mixing 1 g of haematoxylin, 0·2 g of sodium iodate, and 50 g of potassium alum in 1 litre of distilled water. Shake the solution until it is blue-violet, then add 50 g of chloral hydrate and 1 g of citric acid. The solution should now turn red-violet.

i) **Janus green B**. Use this as a very dilute solution, at least 1 in 10,000. Alternatively, a 0·1% solution in alcohol can be applied to the slide and allowed to dry before the water droplet containing the animals is added.

j) **Lactophenol cotton blue**. Dissolve 20 g of phenol crystals in 20 cm^3 of warm distilled water, then add 20 g of lactic acid, 40 g of glycerine, and 0·05 g of cotton blue (methyl blue).

k) **Leishman's stain**. Dissolve 0·1 g of powdered Leishman's stain in 100 cm^3 of methanol, warming for 15 minutes, with occasional shaking. Allow to stand for 24 hours before use.

l) **Lugol's iodine**. Add 4 g of iodine and 6 g of potassium iodide to 100 cm^3 of distilled water.

m) **Methylene blue**. Mix 50 cm^3 of 0·1% methylene blue (in water) with 50 cm^3 of absolute methanol.

n) **Neutral red**. Dissolve 0·1 g of neutral red in 100 cm^3 of 20% alcohol. Filter when dissolved.

o) **Phloroglucin**. Dissolve 1 g of phloroglucin in 100 cm^3 of water.

p) **Safranin** and **light green**. Dissolve 1 g of safranin in 100 cm^3 of 70% alcohol. For light green stain (fast green) dissolve 1 g of light green in 100 cm^3 of clove oil, heat in a water bath, and filter.

q) **Schultze's solution**. Add 50 g of zinc chloride and 16 g of potassium iodide to 17 cm^3 of distilled water to dissolve. Add an excess of iodine and leave to stand for a few days. Decant the stain into brown bottles.

Fixatives

Most fixatives are highly toxic and are therefore difficult to obtain. The two suggested here and the two in Chapter 4 are formulated from more available materials.

a) **Formol saline**. Mix 10 cm^3 formalin (formalin is a 40% solution of formaldehyde in water) with 7 cm^3 of 10% sodium chloride solution and 83 cm^3 of distilled water. For tissues of marine animals use seawater instead of distilled water.

b) **Formalin-aceto-alcohol**. Mix 5 cm^3 of formalin with 5 cm^3 of glacial acetic acid and 90 cm^3 of 50% or 70% alcohol.

c) **Acetic alcohol**. A mixture of 3 parts absolute alcohol and 1 part glacial acetic acid.

d) **Acid alcohol**. 1 cm^3 of glacial acetic acid dissolved in 100 cm^3 of 70% alcohol.

Adhesives

a) **Gum tragacanth** (or similar gum), suitable for dry mounts of crystals, and biological materials such as hairs, butterfly wing scales, etc.
b) **Mayer's albumen**, for sticking sections to slides. Make by mixing equal volumes of egg white and glycerine, then filter. Preserve, if necessary, by adding a few crystals of thymol.

Clearing Agents

Traditionally **xylene** (xylol) is used with Canada balsam, but clove oil, cedarwood oil or pink paraffin can also be used, and may be specified in some cases. See text for use.

Other Chemicals

a) **Alcohol**. All references to alcohol in the text refer to ethanol (ethyl-alcohol), which is usually only obtainable if previously rendered unfit for human consumption, for example by the addition of methanol (methyl-alcohol). For many uses, for instance as a dehydrating medium, acetone may be used instead, to the same dilutions. A further alternative is iso-propanol (iso-propyl alcohol) which is more readily available than ethanol.
b) **Acids**. All procedures described in the text involving the use of acids require only dilute solutions (3–5%), although even at these dilutions the acids should be treated with caution.
c) **Methyl cellulose**. Wallpaper paste is methyl cellulose, and may be used in dilute solution (2% in water) to slow down protozoa. Ensure that the paste is free from added fungicides.
d) **Potassium hydroxide** (KOH). For softening insect parts, use as a 10% solution in water. Sodium hydroxide is equally suitable at the same dilution.

Culture Methods

Only three examples are given here; further culture methods should be sought in appropriate reference books (see Bibliography).
a) *Paramecium* culture. A simple culture method is as follows. Sterilise a large jar by boiling it in water. Add cold boiled water and a dozen rice grains to promote the growth of bacteria as food. Add the purchased culture and cover with self-clinging cellophane. Add six rice grains every month and remove the old ones.

b) *Amoeba* culture. Water of a particular chemical concentration is required. Make up three stock solutions as follows:

i) Dissolve 0·433 g of calcium chloride and 0·162 g of potassium chloride in 100 cm^3 of distilled water.
ii) Dissolve 0·512 g of potassium hypophosphate in 100 cm^3 of distilled water.
iii) Dissolve 0·28 g of magnesium sulphate in 100 cm^3 of distilled water.

To 997 cm^3 of distilled water, add 1 cm^3 from each of the three stock solutions. Pour some of this into a wide glass container with four unboiled rice grains. Leave a few days and then add a *Chilomonas* culture to supplement the bacteria present as food for the amoebae. Add the amoebae, cover with self-clinging cellophane, and store away from the sun. Every two months add four more rice grains and remove the old ones.
c) *Hydra* culture. Keep in a cool place in a jar half-filled with water. Feed with *Daphnia*, or if these are unavailable, with brine shrimp larvae. Hatch out the brine shrimp eggs in salt solution in a warm place, strain them from the water, wash them in pond water and add to the *Hydra* culture. Remove all waste material from the culture.

Appendix II: Suppliers

The suppliers listed deal with individuals as well as with teaching and research establishments, although supply of some hazardous chemicals may not be possible (see Appendix I). For further information, catalogues, etc, individuals should deal directly with the companies. For additional suppliers, consult local microscopical societies (see Bibliography).

Materials available from Drugstores

The following materials are normally held in stock by most drugstores:

glycerine
acetone
clove oil
formalin solution/formaldehyde
citric acid
iodine
chloral hydrate*

*Prescription only.

Most drugstores can easily obtain the following materials from wholesale suppliers:

potassium iodide
phenol
glacial acetic acid
alum potash
absolute alcohol (Graves alcohol) is also available at package stores
paraffin wax is available at supermarkets
dilute nitric and hydrochloric acid may be ordered from the companies listed under General Suppliers

General Suppliers

Bausch & Lomb Co., Scientific Optical Products Division, 420 Route 46, Fairfield, NJ 07006, (201) 575-1646.

Carolina Biological Supply Co., Burlington, NC 27215.

Central Scientific Company, 11222 Melrose Ave., Franklin Park, IL 60131, (312) 451-0150.

Connecticut Valley Biological Supply Co., Inc., P.O. Box 326, Southampton, MA 01073, (413) 527-4030

Creative Publications, P.O. Box 10328, Palo Alto, CA 94303, (415) 968-3977.

Edmund Scientific Supply Co., 101 East Gloucester Pike, Barrington, NJ 08007, (609) 547-3488.

Film Rental Center of Syracuse University, 1455 E. Colvin St., Syracuse, NY 13210, (315) 479-6631.

Fisher Scientific Supply, Educational Materials Division, 4901 W. Le Moyne Ave., Chicago, IL 60651, (312) 378-7770.

Kemtec Educational Corp., P.O. Box 57, Kensington, Maryland 20795, (301) 946-8676.

Kramer Scientific Corp., One Odell Plaza, Yonkers, NY 10701, (914) 476-8700.

Markson Science Inc., P.O. Box 767, Del Mar, CA 92014, (714) 755-6655.

Nasco Science, 901 Janesville Avenue, Fort Atkinson, WI 53538, (414) 563-2446.

Nasco West, 1524 Princeton Ave., Modesto, CA 95352, (209) 529-6957.

Science Kit Inc., 777 East Park Drive, Tonawanda, NY 14150, (716) 874-6020, TWX 710-262-1265.

Science Software Systems, Inc., Audio Visual Products in Science, 11899 West Pico Blvd., West Los Angeles, CA 90064, (213) 477-8541.

Ward's Natural Science Establishment, P.O. Box 1712, Rochester, NY 14603, (716) 467-8400.

Bibliography

Selected references for further reading are listed here. The list is by no means exhaustive but can be regarded as a starting point for obtaining more information, about both microscopy and the specimens examined.

General Texts

Casartelli, J. D. *Microscopy for Students*, McGraw Hill, London and New York, 1965.

Healey, P. *Microscopes and Microscopic Life* (2nd Edn), Hamlyn, London, 1979.

Chapter 1

Barer, R. *Lecture Notes on the Use of the Microscope*, Blackwell Scientific Publications, Oxford, 1953.

Chapters 3 and 4

Bradbury, S. *Peacock's Elementary Microtechnique* (4th Edn), Edward Arnold, London, 1973.

Freeman, W. H. and Bracegirdle, B. *Atlas of Histology*, London, 1967.

Gillies, R. R. and Dodds, T. C. *Bacteriology Illustrated* (4th Edn), Churchill Livingstone, Edinburgh and London, 1976.

Grimstone, A. V. and Skaer, R. J. *A Guidebook to Microscopical Methods*, Cambridge University Press, Cambridge, 1972.

Gurr, E. *Biological Staining Methods* (9th Edn), Hopkins and Williams, Romford, 1978. (See list of suppliers for address.)

Hole, C. B. *An Introduction to Cell Biology*, Macmillan, London, 1974.

NBS Booklets. A series of booklets, written by J. Marson, on various

microscopical techniques. Obtainable from Northern Biological Supplies (see Appendix II).
Purvis, M. J., Collier, D. C. and Walls, D. *Laboratory Techniques in Botany* (2nd Edn), Butterworths, London, 1966.
Shaw, A. C., Lazell, S. K. and Foster, G. N. *Photomicrographs of the Flowering Plant*, Longmans, Green, London, 1965.
Swanson, C. P., Merz, T. and Young, W. J. *Cytogenetics*, Prentice Hall, Englewood Cliffs, 1967.

Chapter 5

Clegg, J. *Freshwater Life* (3rd Edn), Warne, London, 1974.
Clegg, J. *Observer's Book of Pond Life* (3rd Edn), Warne, London, 1980.
Needham, J. G. and Needham, P. R. *A Guide to the Study of Freshwater Biology*, Holden-Day in San Francisco, Constable in London, 1962.
Ward, H. and Whipple, G. *Freshwater Biology*, John Wiley, New York, 1959.

STANDARD TEXTS ON MARINE PLANKTON

Hardy, A. *The Open Sea: the World of Plankton*, Collins (New Naturalist Series), London, 1956.
Newell, G. and Newell, R. *Marine Plankton: a Practical Guide* (5th Edn), Hutchinson, London, 1977.

FRESHWATER ORGANISMS

Identification manuals for freshwater organisms are published as booklets by several institutions,* notably:

Field Studies Council, Nettlecombe Court, Williton, Taunton, Somerset. (Offprints of articles published in their journal *Field Studies*.)
Freshwater Biological Association, Ferry House, Far Sawrey, Ambleside, Cumbria. (FBA Scientific Publications.)
Institute of Terrestrial Ecology, 36 Storey's Way, Cambridge. (I.T.E. Publications.)

Algae

Belcher, H. and Swale, E. *A Beginner's Guide to Freshwater Algae*, I.T.E. Publication, H.M.S.O., London, 1976.

* Lists of publications may be obtained from these by sending s.a.e. to the addresses given.

Belcher, H. and Swale, E. *An Illustrated Guide to River Phytoplankton*, I.T.E. Publication, H.M.S.O., London, 1979.
Lund, J. *The Algae of the Malham Tarn District*, Field Studies Offprint, 1961.

Protozoa

Corbett, S. *Microscopical Animals in Sphagnum Moss: Illustrated introduction to the testate rhizopods around Malham, Yorkshire*, Field Studies Offprint, 1973.
Curds, C. *An Illustrated Key to the British Freshwater Ciliated Protozoa Commonly Found in Activated Sludge*, Water Pollution Research Technical Paper No. 12, H.M.S.O., London, 1969.
Page, F. C. *An Illustrated Key to the Freshwater and Soil Amoebae*, FBA Scientific Publication No. 34, 1976.

Culture methods and preparation methods for protozoa are given in:

Mackinnon, D. L. and Hawes, R. S. J. *An Introduction to the Study of Protozoa*, Clarendon Press, Oxford, 1961.
Sleigh, M. *The Biology of Protozoa*, Edward Arnold, London, 1973.

Hydras

There are no easily obtainable booklets on hydras, but the following technical publications should be consulted:

Forrest, H. *Taxonomic Studies on the Hydras of North America, Part 8*, Transactions of the American Microscopical Society, vol. 82, pp. 6–17, 1963.
Grayson, R. F. *The Freshwater Hydras of Europe. 1. Review of the European Species*, Archive fur Hydrobiologie, vol. 68, pp. 436–449, 1971.
Maxwell, T. *The Freshwater Hydras of Europe. 2. Description of Hydra graysoni sp. nov.*, Archive fur Hydrobiologie, vol. 69, pp. 547–556, 1972.

Crustacea

Harding, J. and Smith, W. *A Key to the British Freshwater Cyclopid and Calanoid Copepods*, FBA Scientific Publication No. 18, 1960.
Scourfield, D. and Harding, J. *A Key to the British Species of Cladocera – with notes on their ecology*, FBA Scientific Publication No. 5, 1966.

Others

Morgan, C. and King, P. *British Tardigrades*, Academic Press, London, 1976.
Mundy, S. *A Key to the British and European Freshwater Bryozoans*, FBA Scientific Publication No. 41, 1980.

Chapter 6

Allman, M. and Lawrence, D. *Geological Laboratory Techniques*, Blandford Press, Poole, Dorset, 1972.

Hallimond, A. F. *The Polarizing Microscope* (3rd Edn), Vickers Instruments, York, 1970.

Smith, H. G. *Minerals and the Microscope* (4th Edn – revised by M. K. Wells), Allen and Unwin, London, 1956.

Chapter 7

Hawkins, A. and Avon, D. *Photography: the guide to technique*, Blandford Press, Poole, Dorset, 1979.

Kodak Ltd. *Photography through the Microscope* (6th Edn), Eastman Kodak Co., Rochester, New York, 1974.

Walker, M. I. *Amateur Photomicrography*, Focal Press, London, 1971.

Chapter 8

Bradbury, S. *The Optical Microscope in Biology*, Edward Arnold, London, 1976.

Burrells, W. *Microscope Technique* (2nd Edn), Fountain Press, London, 1971.

Chapters 9 and 10

Grimstone, A. V. *The Electron Microscope in Biology* (3rd Edn), Edward Arnold, London, 1976.

Meek, G. A. *Practical Electron Microscopy for Biologists* (2nd Edn), Wiley-Interscience, 1976.

Microscopical Societies

The following societies will provide advice and information on microscopical technique to their members:

United Kingdom

Liverpool Microscopical Society: Secretary, M. L. Bellis, 5 Easton Road, New Ferry, Wirral, Merseyside.

Manchester Microscopical Society: Secretary, J. Lester, 12 Buckland Avenue, Blackley, Manchester.

Postal Microscopical Society: Secretary, Barry Ellam, 117 Straight Road, Lexdon, Colchester, Essex. (Circulates boxes of slides with notes to its members.)
Queckett Microscopical Club: % British Museum (Natural History), London.
Reading Microscopical Society: Secretary, P. G. Vidgen, Forest Keep, Watersplash Lane, Ascot, Berkshire.
Royal Microscopical Society: Secretary, Dr A. W. Robards, 37–38 St Clements, Oxford.
Freshwater Biological Association: Secretary, The Librarian, Ferry House, Far Sawrey, Ambleside, Cumbria. (The association fosters freshwater microscopy.)

Australia

Australian Microscopical Society: Secretary, R. H. Freere, Wild Australia Pty Ltd, Sydney, New South Wales.

Canada

The Microscopical Society of Canada: 150 College Street, University of Toronto, Toronto, Ontario, MS5 1A1.

United States

New York Microscopical Society: American Museum of Natural History, Central Park West at 79th Street, New York, NY 10024.

Photographic Credits

We would like to thank the following for permission to use the photographs listed:

Dr R. Butler (figs. 56b and 136b). Cambridge Instruments Ltd (fig. 138). Gallenkamp & Co. Ltd (figs. 6, 7 and 93). Dr M. Harris (plate 1). Jeol (UK) Ltd (fig. 123). Kratos Scientific Instruments (fig. 127). E. Leitz (Instruments) Ltd (fig. 47). Mr C. Lister (plate 6). Mr P. Matthews (figs. 62 and 109). Reichert-Jung UK (fig. 120). Dr D. Sigee (fig. 116). Miss J. Roberts (figs. 129 and 130). Vickers Instruments (fig. 90). All other photographs by Alan Curry.

Index